HIRE • TRAIN • RETAIN

"Over the first ten years of our entrepreneurial careers, we agonized over the hiring process. Without any formal training in human resources, our lackluster results proved we needed help!

"For the past several years, Michael Matalone has mentored our staff and the two of us on the XP3 Talent System with overwhelming success. Of greater importance, he equipped and empowered our team to do it themselves! Over and over, we would tell Michael, 'You must write a book and share your insights with others.' Finally!!

"In Hire • Train • Retain, you will learn a step-by-step, holistic process for building an 'A Team,' filled with exceptional talent: From how to utilize behavioral profiles and evaluate motivators, to a complete onboarding plan that ensures your new team member is 'wowed' by how *much you care about their success. Do yourself a favor and buy two copies—one for you, and one for your best friend. They need it, too!"*

—Bryan Miller, John Gwaltney, Owners/Partners
Virtus Family of Companies

My only regret is that this book did not exist 30 years ago when I started my career in HR. This is a must read, step-by-step guide to getting your people programs right from hiring to coaching to engagement."

—A.Leigh DeWulf, SHRM-SCP, Founder HR.Coach

Hiring, training and developing people is central to any organization's success, yet it's an area where we have the least amount of knowledge. This book provides an easy to follow systematic way to significantly improve hiring success, increase employee engagement and ultimately lead to greater results. A must read for anyone who hires and manages people.

—Doug Johnston, Author
Thriving in Conflict, and *Exponential Leadership*

HIRE

TRAIN

RETAIN

HIRE

HOW TO GET THE RIGHT PEOPLE

TRAIN

IN THE RIGHT ROLES

RETAIN

DOING THE RIGHT THINGS

Michael Matalone

ISBN: 978-0-578-55143-2 - Paperback

Library of Congress Control Number: 2019918077

Printed in the United States of America   0 2 2 6 2 0

♾ This paper meets the requirements of ANSI/NISO Z39.48-1992 (Permanence of Paper)

To my wife, Eva for her constant encouragement, companionship, and kindness. You are the wind beneath my wings. I love you with all my heart.

CONTENTS

- The science that underpins your future success in hiring, training, and retaining top talent
- Personality traits, behavior, and their sources
- The three foundational strengths of a productive employee
- Psychometrics: What you need to know about personality assessments
- The three primary behaviors
- Introverts, ambiverts, and extroverts
- The vital "AD" and "PRB" scores

FOREWORD

by Serge Beauchemin

Respected entrepreneur, investor, thought-leader, and star of the the Canadian TV equivalent of "Shark Tank."

Without a doubt, one of the most important pillars of success in business is the human component. All organizations that have withstood the test of time can attribute a disproportionate share of their growth and success to the talents of their teams. Indeed, even the best-laid plans will fall short without dynamic individuals on hand to execute them.

When it comes to adding members to your team, we are now faced with a diversity of distinct challenges. There's "generational shock," which happens when Baby Boomers, Gen Xers, and Millennials are mixed together, forced to team with each other, despite their often-widely diverging values and beliefs. Add to this the challenge of multiculturalism: While different customs and cultures certainly enrich the depth of teams, they also create headwinds when it comes to things like collaboration, integration, and communication.

Now more than ever, organizations seeking to succeed—both in growth and sustainability—must master the art of forming motivated teams, mobilizing them toward a common goal, while developing and maintaining top talent.

We live in the information age. Those who know the most about human capital possess an outsized advantage. When it comes to recruitment, all companies, in all industries, are competing to find those rare pearls, those needles in the haystack.

This is hard. To do it, you must become a true talent specialist. You must find those who will complement your teams. You must also excel in the internal development of talent, post-onboarding, and implement foolproof plans to keep them. Indeed, "retention rates" are increasingly among an organization's key performance indicators, alongside "debt ratio" and "return on investment."

But how can you possibly form successful teams in such a setting? How can you identify the talents that help your troops conquer the highest peaks? How can you take advantage of richness and diversity to form competent and complementary teams wherein solidarity and consensus combine to exceed stated goals? How, indeed, can you ensure that you provide everyone with what they need to be truly happy in their roles, and thus live up to their full potential, as individuals and a team?

This book answers all of these fundamental talent-management questions. Backed by decades of experience in talent selection, recruitment, and development, Mike Matalone has created a powerful approach and toolbox for addressing all these challenges. It's the XP3 Talent System. This makes is possible—even easy—to select, train, and retain top talent, according to their knowledge and their know-how—while respecting a role's functional requirements and the dynamics of the larger team.

For years, however, the only way you could gain any insights or skills in the XP3 Talent System was to engage Mike and his team in person. I, and countless others, had been asking—even begging—him to "Please write a book about this!"

That book is in your hands now. I think it captures not only the essential tools, protocols, and techniques you'll need to succeed, but it also captures Mike's unique and engaging personality. It's like you'll have him at your side, every step of the way, making things easy, and ensuring your success at every step.

I hope you enjoy this book. I know I did.

INTRODUCTION
The science of success

The average company spends up to 40 percent of its total revenue on employee compensation. Yet what does it get for that money? Put it this way: *Are you currently maximizing your company's biggest investment?* Do you have a locked-down system for finding, training, and then keeping the world's best employees and managers? Do you know what it takes to get, and hang on to, the invaluable "A Players" who can propel your company to unprecedented growth?

There are lots of ways to approach this challenge; most are difficult to learn and implement—and many simply don't work. Clearly, you want the best tools, knowledge, and skills at your fingertips: a proven system that's easy to learn and use, one that arms you and your team with the essential skills for success.

Only one approach checks all these boxes: The XP3 Talent System. In this book, I'm going to give you the tools, protocols, and shared language you need to literally shave years off your company's profit trajectory. Consider this book an "owner's manual for your people."

Bonus: It's easy. Double-bonus: It's fun.

The numbers prove me out. And so does the science. You want numbers? In just a minute, I'm going to tell you the story of Jake, a guy who couldn't seem to get ahead in the hiring/training/retention game. And you want science? Let me start with a little tidbit that should get you thinking in the right direction—in the *XP3 Talent* direction—of how you have been hard-wired, from birth, to operate in ways that don't always fit the situation at hand.

Let's take hiring. I've worked with scores of hiring managers over the years, and most of them will initially tell me something like, "I've got a good gut feeling when I hire. I can tell within the first five minutes of an interview if that candidate will succeed at the role."

In a word, Ouch. Hiring a person for a specific role is a very specific endeavor. Your gut feelings don't work here. In fact, when you make a "gut call," all you're doing, unconsciously, is comparing that candidate to yourself. That's how our brains are hard-wired. If you're a cautious or analytical thinker, for example, you'll gravitate toward candidates with those same tendencies.

There are two problems with this gut-feel/hard-wired approach. First, *your* behaviors may not be what's required to succeed in the role that you're hiring for. And second, you still need to overcome what the *candidate* is doing during that interview. They're trying their hardest to be the best possible interviewee and impress you and win the job. Is that the same person they'll be, say, three months down the line? Gut feelings don't work here. Science does. And that's what the XP3 Talent System is all about. It's science, made easy, usable, repeatable, and profitable.

And what about Jake, the guy who needed to reduce turnover? Let's find out!

Think big!

Jake (not his real name) was trying to single-handedly run the sales force of his company, a residential services firm that was ranked among the top three in its industry.

Jake told me his problems. He couldn't find enough good people. The ones he found wouldn't stick. The turnover was costing him a fortune.

(According to the U.S. Department of Labor, the price of a bad hire is at least 30 percent of that employee's first-year earnings! For a small company, a five-figure investment in the wrong person is a big gamble that managers make every time they make a hiring decision—only they are betting using "house money." Are you willing to risk that much money without the proper training and coaching to boost your odds of success? These losses add up and can be a real threat to your business.)

Back to Jake. He had built his company to its current impressive size using his gut-feel approach, but now he'd hit a wall. "It's not like there's a science to this," he told me.

"Actually, Jake," I replied. "There is."

I helped Jake bring the XP3 Talent methodology to his company. We built an executive management team. We trained them all on the XP3 Talent protocols. We helped them hire a class of 18 sales trainees—and 17 of them graduated, a record-breaking rate. In fact, their average final exam score was 93, beating the previous average of 81.

But turnover was just the tip of the iceberg. We boosted the *effectiveness* of this graduating sales class, too. Remember I mentioned "numbers"? In just six weeks, we found, onboarded, and trained a new class of sales reps who were able to bring in nearly $1.5 million in new sales their first 30-days in the field. And this was just the first graduating class that we helped to bring on board. We graduated another half-dozen classes afterward, demonstrating the scientific repeatability of the XP3 Talent System.

Ready, aim... don't fire!

Why do people quit jobs? They don't. They quit *managers*. ("I hired the best people, but they all hated working for me!") In other words, you'll be wasting time and money hiring top talent if you don't first ensure that your entire management team is in the right roles, doing the right things.

I once met a manager who told me she would automatically fire two of the new hires in any given class of trainees, *"just to send a message to the others."* It sent a message, all right—but hardly the right one!

The quick lesson here—and we'll get into this in more detail later in the book—is that *"A Players" hire "A Players," and "B Players" hire "C Players."*

Why is that? "A Players" want to work with other "A Players." That's because they realize that, together, they can accomplish a lot more than they can by themselves.

I was in an Apple store recently, and a sharp

young guy was helping me out. He was energetic, proactive, resourceful, attentive to my needs, and very knowledgeable about Apple's products. After he helped me, I asked him why he liked working for Apple. He said, "See this blue shirt I'm wearing? Now look at all the other blue shirts here. Every day I come to work and get to work with very smart people who help me learn and grow, which enables us to have the opportunity to better help people like you." Isn't that the kind of culture you want to foster in your organization?

Now consider the flip side. Most "B Players" hire "C Players," because they're typically threatened by people who are smarter than they are. They're afraid the "A Players" will make them look bad or possibly even take their jobs!

Here's the bottom line: If you're going to hire top talent, then the first thing you need to do is ensure that your team which is tasked with hiring, training, and retention is made up of "A Players." It's a top-down approach.

Overcoming "the hiring bias"

Every company grows or dies based on its key employees' and managers' ability to scale the organization. Every success or failure begins with their abilities to do what's needed to support growth. Market conditions will always change, but a high-performance team will dramatically—and consistently—improve your organizational effectiveness, delivering a more sustainable competitive advantage.

Unfortunately, far too many people believe that success hinges primarily on their ability to simply "find the right person," which is why so many companies are willing to invest so much in recruiting!

But here's the reality: Hiring talented people is not enough to ensure your organization's success. Don't expect them to fly on auto-pilot. You must have engaged, productive employees in order to achieve your growth objectives—and this is a complex and *ongoing process* that requires your constant focus and action. If you want to be a market leader, you need to invest in developing your managers, and the infrastructure required to support their growth… enabling and empowering them to evolve the professional skills required to ensure your company's continual success.

In other words, you'll want to close "the talent gap": the void that lies between your company's growth or performance objectives, and your staff's ability to execute consistently to achieve the required results within the timeframe that you define.

For example, if your growth plan is to double your company's revenues and profits in the next three years, then you don't have time to wait for the learning curve—which research has shown can last anywhere from three to five years—of a new or inexperienced person in any critical role. To the contrary: You'll want to ensure that you have top talent in place who can hit the ground sprinting, and immediately execute on your growth plan.

With this understanding, you have two options at your disposal:

1. **You can lower your expectations** and compromise on your strategy to fit the talent capabilities of your existing people and the new unqualified people you will hire. This translates to lost time, money, and reduced odds of achieving your goals.
2. **You can improve your talent capabilities**, which includes assessing your existing staff's abilities; finding new, highly-qualified talent; and then constantly developing your teams' talent to ensure they possess the latest, greatest knowledge, skills, and abilities that are required to achieve your growth objectives within your desired timeframe.

Put in this light, the choice seems like a no-brainer. But it can be devilishly hard to perceive. Most business owners don't even realize that they have a talent gap in the first place. Stick with me, and I'll teach you how to identify it—and close it.

Why I wrote this book

The proprietary XP3 Talent System has been scientifically designed and honed over the years to make it the easiest, most effective, and foolproof way to turbocharge business success along the essential lines of hiring, training, and retention. It's been used to help thousands of people across scores of companies in countless industries. Previously, it's only been available by participating in the XP3 Talent Leadership Academy's intensive live workshops. But more and more people kept asking me, "Couldn't you put this into a book?" Well, I did, and this is it. It's your way of getting a jump-start on all this knowledge and power, in one place, quickly and easily.

I'm not exaggerating when I say "quickly and easily."

This book is made of meat, not fat. I'll share a story where it helps to convey a concept or make a point, but I will not bog you down in fluff. I don't want you to waste a minute of your valuable time.

And is it easy? Let me ask you this: Can you count to three?

Of course you can. And that's the "3" of "XP3." Wherever possible, the rules and techniques in this book harness and leverage that simple, magic number that you recognize from countless other easy-to-remember things, such as:

- Body, mind, soul
- The Olympics: Gold, Silver, and Bronze medals
- The primary colors: red, yellow, and blue
- A traffic light: Stop, Caution, and Go
- Geometry: The three-sided triangle is nature's most stable shape

Move this to business, and you'll recognize these "three's":

- Sales leads: Hot, Warm, and Cold
- Employee Performance: A, B, or C Players

Bring it to the realm of the XP3 Talent System, and you'll find even more to the Power of Three:

- The three keys of the world's greatest managers
- The three strengths that create a consistent, productive, and engaged employee

- The three primary behaviors and personality traits
- The three-step interview methodology

And the XP3 Talent System builds upon itself. No concept exists in a vacuum. So what you'll learn about hiring will apply to training… which will apply to retention. It actually gets easier as you go! There are a lot of methodologies out there for what this book covers, but I promise you, none are so easy to grasp and utilize. You're going to have a lot of fun as you succeed.

Everything you need

As I'd mentioned above, the XP3 Talent System was previously available only to sponsors and participants in the XP3 Talent Leadership Academy. This book is derived from—and expands upon—that material. It includes:

- The XP3 Talent Applicant Screening Questionnaire.
- The XP3 Peak Performance Profile (P3) job description.
- The XP3 New Hire Onboarding Training Program.
- The XP3 Talent Career Progression Skills & Assessment Tool.

- The XP3 Talent Performance Coaching Guide.
- And a lot more!

Are you ready to take your company to a whole new level? Are you ready to build and lead a truly world-class team?

I know you are! Let's get started! It's my pleasure to be your guide.

CHAPTER 1

The science of talent management

What you will learn in this chapter:

- The science that underpins your future success in hiring, training, and retaining top talent
- Personality traits, behavior, and their sources
- The three foundational strengths of a productive employee
- Psychometrics: What you need to know about personality assessments
- The three primary behaviors
- Introverts, ambiverts, and extroverts
- The vital "AD" and "PRB" scores

The science of understanding people

Businesses today pride themselves on their use of data. Those that are really good at it, exploit what's known as "Big Data."

Yet far too few businesses use data to manage their single most critical resource: Their people.

That's a challenge which the XP3 Talent System is going to help you, and your company, to overcome. This chapter will give you the scientific foundation upon which the subsequent chapters—devoted to hiring, training, and retention—will build. Think of it as a holistic approach to understanding people, and what makes them tick.

In this chapter, I'm going to give you scientific insights, and tools, to improve the decisions you make and sharpen your skills as a leader and manager. These skills only grow in importance as the complexity and value of the subject grows; that's because the role of the data itself changes. At the most basic level, all you need is raw data. But as it grows, you'll need to be able to extract insights from it in order for it to be useful to you; it's the difference between, say, a flat drawing and a three-dimensional model. As the data continues to grow in complexity (and value), you'll then need to take your newfound insights and be prescriptive about which actions you should be taking.

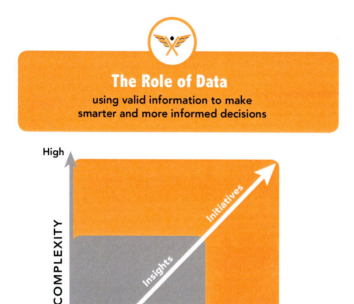

The Role of Data

using valid information to make
smarter and more informed decisions

In other words, in order for you to be a highly effective manager—to excel at applying the scientific principles of hiring, training, and retaining top talent—you must first develop a basic understanding of the underpinnings of human behavior, and their impact on employee performance.

What are personality traits?

Personality traits are inborn. They're what make you, you. They begin as links between brain cells (a.k.a. neurons) known as synaptic connections or neural pathways. The forging of these pathways begins about 60 to 90 days prior to birth, and continues until around age 15.

These synaptic connections form the basis of how you think and feel; therefore, they determine how you will behave in most situations.

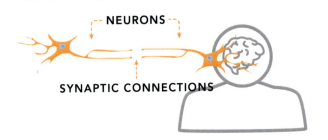

The Anatomy of Your Traits

NEURONS

SYNAPTIC CONNECTIONS

Of course, not everything you do is pre-programmed. The way you were raised, educated, and mentored—by your parents, relatives, friends, teachers, coaches, and others—impacts the "programming" of your "hard-wired" way of thinking. This, too, affects how you will behave in various situations.

How does this translate our discussion of workplace behavior? Consider this: The more "hard-wired" your way of thinking becomes, the more rigidly you define "normal" behavior. That, in turn, will affect the degree of conviction you will have in asserting your beliefs to others, regarding what you consider to be "right" or "wrong" behavior.

Intrinsic vs. extrinsic personality factors— and how they affect behavior

The hard-wired factors we discussed above are *intrinsic*; these account for our predictable behaviors, based on our internal needs and drives. They're a reflection of our core personality traits.

Extrinsic factors—such as your education, experience, the situation, the resources at your disposal, your values, etc.—combine with their intrinsic counterparts to account for your overall behavior. Yet the *balance* of intrinsic vs. extrinsic has a profound affect on your performance.

Doing work that is intrinsic to us gives us physical

energy. Doing work that we *have to do*, but would prefer *not* to do (i.e., work that is *extrinsic* to us), *consumes* energy. We all have to do both—but if you spend more time using your *intrinsic* behaviors, you will be more productive, engaged, and leave work with more physical energy.

Intrinsic vs. Extrinsic
The Difference

Intrinsic

Extrinsic

Consider the following example:

You come to work on Monday to discover a towering stack of to-do items atop your desk. Given the option, you'll always cherry-pick the ones that are the easiest, most appealing, and sometimes even fun to do. As you keep checking things off the list, the pile gets smaller. At the end of the day, you'll know you've accomplished a lot, and will feel energized for having done it.

But then it's Tuesday. You return to work to see all the things on your desk that you opted to avoid on Monday. You still need to complete them, so you hunker down and start plowing your way through them. One by one, you're marking to-do's as "Done," but each one saps you of energy. It's nothing like Monday. By the end of the day, you're exhausted.

The following graph summarizes the impact of behavior on performance. Remember I mentioned earlier that this chapter will be providing you with "tools" to put behavioral science to work for you? Well, the main tool is what's called a "personality assessment." We'll be getting into that in detail later in this chapter; in the meantime, just look at how this graphic shows that, once you have insight into an individual's perception and interpretation of a situation, you can use it to predict their behavior and impacts results.

Behavior Origin and Impact

INDIVIDUAL
Personality, Traits, Knowledge, Experience, Motivators, Values, Needs

X

SITUATION
Job, Corporate, Organization, Culture, Events, Communication, Manager

=

BEHAVIOR
Actions, Reactions, Verbal, Expression, Relationship, Attitudes, Decisions

X

RESOURCES
Product, Management, Support, Capital, Equipment, Structure, Processes, Training

=

RESULTS
Quality, Quantity, Time, Service, Intentional Culture, Profit

What is behavior?

Everything we do is related to our desire to satisfy a fundamental need. And behavior is what we do. It's defined as an action or reaction to your way of thinking and feeling. Learned behaviors are based on your traits, knowledge, values, and motivation.

And they're relatively flexible: With knowledge, experience, and motivation, you can adjust your behaviors to accommodate a given situation. Unlike personality traits that are hard-wired, behavior is mostly learned, but the way it manifests itself is strongly linked to *just how hard-wired* your personality traits are.

The three foundational strengths of a productive employee

Remember, in the introductory chapter, how we talked about the power of three, and how easy it is to grasp new concepts that use it? Here's a great example.

While there are many strengths that people bring to a job, research (by Gallup, Behavioral Science, and others) has shown the value of the following three building blocks. These foundational elements are what enable an employee to perform at a high level of consistency. They're also what allow the employee to *enjoy* the work they're doing (at an intrinsic level).

This happiness goes hand-in-hand with the crucial level of *engagement* that we will discuss in chapter 3. A truly engaged employee does better work: he or she will go above and beyond what's merely required. An engaged employee is a company advocate, spearheading its success. And it almost goes without saying that an engaged employee will stick around; you don't even have to get to the "Retention" chapter of this book to figure that out.

The 3 Foundational Strengths

that create **consistent performance**
and an **engaged happy employee**

RESULTS

Define how success be measured in the role. Then, the application of the below knowledge; when applied appropriately, will create these desired results.

KNOWLEDGE

This refers to knowledge that's specific to the role's requirements—in contrast to, say, advanced degrees that have nothing to do with the role. Examples may include intermediate or advanced skills with MS Word, Excel, QuickBooks, a CRM system, or other software applications. It could be an understanding of a structured sales process, product knowledge, or how to develop a website, etc.

BEHAVIOR

This component is the essential *alignment* of the required behaviors for the role and the hard-wired personality traits of the individual doing the job or being considered for the role. This alignment—and identification of the traits to be aligned—will be determined by the personality-assessment tool, which I'd mentioned earlier, and which we'll explore in greater detail shortly.

Psychometric instruments

Well, I've been teasing you about "personality assessments" all through this chapter; here's the section where your patience will be rewarded.

Personality assessments come to us from a branch of science called psychometrics. Psychometrics studies the measurement techniques practiced in psychology, as well as techniques for validating these measures. High-quality *psychometric instruments* can measure the various aspects of an individual's personality.

Psychometric instruments are more commonly known as "personality assessments." They're also one of the most important tools for helping you to hire, train, and retain top talent.

"Personality assessment." "Psychometric instrument." "Validation of psychological measurements." It all sounds very scary and intimidating. But don't worry. While there certainly is a lot of hard and wide-ranging science behind all these big words, I'm going to filter out all the stuff you don't need to worry about. I'm going to distill it down to just the parts that matter. And I'm going to walk you, slowly and carefully, through all of these new concepts and tools.

Choosing the best psychometric instrument

First things first: A *psychometric instrument is not a test!* Don't ever refer to it as a "test," because it isn't one. Most people already put enough pressure on themselves, so there's no need to add to that stress by telling someone you're going to "test" them! Over-stressing the candidate, incidentally, can negatively impact your ability to get a valid result. So remember: It's not a "test"!

That said, how do you choose the psychometric instrument—a.k.a. personality assessment—that you, as a manger, will use? There are hundreds on the market today.

But some are better—vastly better—than others. The best ones will have the following characteristics and benefits:

1. They are easy to learn.
2. They are easy to administer.
3. They are quick; it shouldn't take your candidate more than about 10 to 15 minutes to complete one.
4. They offer multiple tools to help. At a minimum, they should include a "job profiling" tool that helps determine what

specific behavioral traits each of your company's roles requires to be successful—and then an individual assessment that is then aligned and compared to determine the fits and gaps.

5. They are current and include the latest advancement in psychometrics. Science and technology are always advancing; therefore, a valid instrument is continually being updated to incorporate these advances.

6. They are affordable. This means that "cost" should never be a consideration as to whether you will administer one. Generally, you'll want to avoid a per-test cost structure (a disincentive for using them), and look instead for a blanket annual license—which actually incentivizes you to administer as many assessments as possible.

Introducing the MPO

I've been trained and certified in a variety of these tools since 2002, and for nearly ten years, now, based on the above criteria, I have become an advocate of an assessment known as the "MPO"—an acronym for Managing Performance in Organizations.

The MPO is offered by a Canadian company called Ngenio (www.Ngenioworld.com), and it meets all of the above criteria—and more. Specifically, the

MPO program will effectively enhance your talent acquisition, organizational development, and human resource management strategies.

The MPO program is founded upon a scientifically developed personality inventory and an exclusive job satisfaction survey. The MPO was designed and constructed by standards of the American Psychology Association (APA), the Canadian Psychology Association (CPS), and the U.S. Equal Employment Opportunity Commission (EEOC), a federal agency that administers and enforces civil rights laws against workplace discrimination.

A WORD FROM NGENIO, THE MAKER OF THE MPO

Ngenio, Inc. is happy to contribute to this book by providing information on our MPO program. No doubt, the strategic information provided to MPO-trained managers can greatly assist in building very productive and profitable organizations.

The technical information and explanation on MPO measures offered in this book only scrape the surface—they're obviously insufficient, on their own, to make effective employee decisions, as they represent a small percentage of the information provided in the full MPO training program.

The full MPO Management Training program is a practical two-day training session that can be completed on its own, or as part of the 40-hour, five-day XP3 Leadership Academy.

Ngenio cannot be held responsible for decisions made without good judgement and full information on the people and situations involved.

My company, XP3 Talent, is one of the original distributors and authorized trainers of the MPO in the United States. To learn more, contact us at Info@XP3Talent.com.

A multi-purpose tool

The MPO is like a Swiss Army knife of psychometric instruments. You can use it for:

- Recruiting and selecting candidates
- Leadership development
- Consolidating and mobilizing teams
- Performance management

- Coaching
- Conflict resolution
- Identification of key competencies
- Career mapping and succession planning
- Organizational design

From one single candidate assessment, administered in just minutes, the MPO can produce three different reports and also includes a variety of other tools. These include:

1. **The MPO Job Profile.** This helps you identify the required behavior for a role. You can then use that required behavior as a benchmark by which you can measure the individual's MPO Personality Profile (next on this list) to gauge its fit to the position (using the Right Match report, Number 5 on this list).

2. **MPO Personality Profile.** This measures various facets of an individual's personality to help you get the right people in the right roles, and how to manage them based on their unique needs. (This includes the vital PRB, or Perception of Required Behavior score, which we'll be talking about in detail, later in this chapter.)

3. **MPO Talent.** This highlights a person's potential within a particular sphere and in line with sought competencies. It

provides a thorough report, encompassing 46 of the most crucial talents for the workplace.

4. **MPO Communication.** This provides an overview of an individual's personal communication style. It will help you prevent and manage interpersonal conflict—and coach them on improving team relationships.

5. **The MPO Right Match Analysis Report.** This helps you determine how good of a fit the person is for a particular role by analyzing the MPO Job Profile against the individual's MPO Personality Profile.

6. **MPO Multigraph.** This helps you view and evaluate multiple individuals' MPO Personality Profiles and build high-performing teams.

The MPO graph and how to read it

Quick: How do you gauge the health of someone's heart? You don't have to be a doctor to know that an EKG, or electrocardiogram, provides a fast visual readout of essential factors, such as heart rate and rhythm. This provides medical personnel with the vital information they need to make faster and better decisions regarding the treatment of their patient.

Why does MPO use a graph?

Heart

Electrocardiogram

Personality Profile

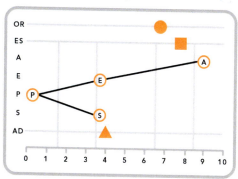

The MPO is similar. It provides a "readout" of the individual's assessment on a graph. This tells the person reading the graph (that is, *you*) what's going on in that person's brain. Specifically, the MPO reveals how the person is "wired" to think and feel; this, as we've discussed earlier, predicts how they will behave in most situations.

Once you have this information in hand, you can make faster and better decisions regarding hiring, promotions, how best to train, communicate, motivate, and coach individuals. That's because you'll be able to do it all, based on their unique personality traits, behavior, and intrinsic needs.

Traits measured by the MPO

The MPO measures a variety of different traits. Each is measured on a sliding scale, and is coded as one- or two-letter abbreviation, specifically:

OR: Originality of thought
ES: Emotional **S**tability
A: Assertiveness
E: Extroversion
P: Pace
S: Structure
AD: Adaptability

Now look at the chart below; it shows you the extremes for each of these traits; for example, the "E" score can vary between "Introversion," at one end of the scale, and (not surprisingly) "Extroversion" at the other:

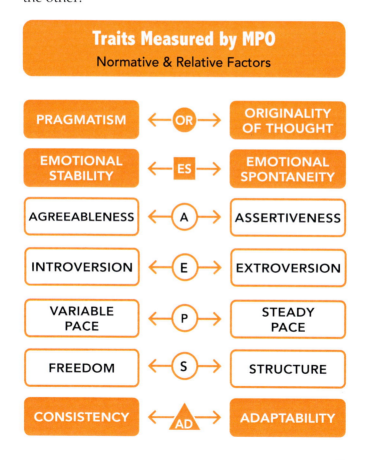

The three primary behaviors

Using the MPO, you can get a better handle—an "inside view"—of an individual's three primary behaviors and traits. These are:

1. **Thinking and decision-making:** How that person scores between "Big-Picture/Strategic" and "Detail/Tactical."
2. **Social orientation:** Whether that person is more at home among "People" or "Things."
3. **Work pace:** How fast and urgent, or slow and methodical, they are.

How to read the graph: An example

Take a look at the sample graph that follows. On the left side, you'll see four of the factors we'd just discussed only this shows the High and Low sides, specifically:

A= Agreeableness or Assertiveness
E = Introversion or Extroversion
P = Fast Pace or Slower Pace
S = Freedom from Structure or Structured

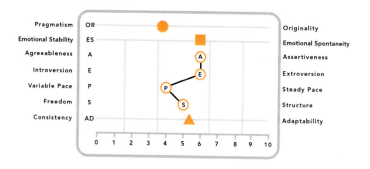

It's the *combination* of these factors that will give you insight into how the individual is wired.

For the three primary traits, for example, here's how they fall:

1. For **thinking and decision-making,** you'll compare "A" ("Assertiveness") and "S" ("Structure") scores.

2. For **social orientation,** you'll compare "E" ("Extroversion," in this case, "People") and "S" ("Structure," in this case, "Things") scores.

3. For **pace,** you'll compare "A" ("Assertiveness") and P ("Pace") scores.

Now consider the following specific combinations, and *what they tell you about how that individual is wired*:

THINKING AND DECISION-MAKING:

- If **A>S**, this person is "Strategic/Big Picture." For them, "close" is good-enough thinking.
- If **S>A**, this person is "Tactical/Precise." They employ detailed thinking.

SOCIAL ORIENTATION:

- If **E>S**, this person is "Extroverted." They're quick to trust, and optimistic. They like, or even need, to work with others to get energized.
- If **S>E**, this person is "Introverted." They're slow to trust, cautious, and exhibit analytical thinking. They need quiet time to get energized, since too much interaction with others may drain their energy.

PACE:

- If **A>P**, this person is "Impatient." They're fast-paced, exhibit a sense of urgency, and get bored with routine or frustrated by a slow pace.

- **If P>A**, this person is "Patient." They operate at a slower pace. They're methodical, and process-orientated. They get frustrated when the pace moves too quickly, and may lack a sense of urgency.

Pretty powerful revelations from a ten-minute assessment, wouldn't you say?

(Incidentally, this book is structured to let you read, and benefit from, the information you'll get from the MPO. But if you hunger for absolute mastery of these skills, you can go beyond what this book offers by investing in a two-day MPO Management Certification Training Program. Learn more by visiting XP3Talent.com.)

Introverts, ambiverts, and extroverts

I think you already know what extroverts and introverts are. The former are outgoing "people-people"; the latter are more reserved.

You may have never heard the term "ambivert" before, but I'll bet you can guess its meaning, simply from the context above. An ambivert is someone who falls between those two other extremes, with a little of each personality trait. *Where* they fall on this continuum, you'll recall, is measured by the "Social orientation" parameters ("S" and "A") of the MPO. It's one of the most commonly-used—and useful—features of the MPO.

How do these different personality types perform in different roles? And why should you, as a manger, care? Let's use the role of "salesperson" as an example. I think you'll find the following stories interesting and illuminating.

EXTROVERTS

We have all been led to believe that the best salespeople are extroverts. However, extensive studies (the Objective Management Group's Modern Science of Salespeople Selection and others) have proven that this is not exactly true. While extroverts do enjoy working with other people, they also have a *need to be liked* by other people; this actually puts them at a great disadvantage in a sales role.

For example, when you have a great need to be liked by other people, you will not handle rejection very well. In fact, you'll take it personally. It wasn't that the prospect didn't like your product, service, or company; the extrovert interprets this as "They didn't like *me*." They then carry that baggage with them to the next sales call, and the next, greatly impacting their ability to close any deals.

Another example of behavior that results from this

"need to be liked" is what's known as "show up and throw up syndrome." Extroverts believe they have to "persuade" people to like them—and this results in their talking themselves out of the sale. It comes across as "Let me tell you all about our products and services and why you should buy from us, etc., etc." That's the exact opposite of asking good questions about what the prospects need, and then listening and learning so you can respond appropriately.

Therefore, to be effective in sales, extroverts must learn to adjust their natural way of thinking and feeling so that they can modify their behavior to be effective. For some, this is a very difficult thing to sustain.

INTROVERTS

Introverts, on the other hand, if too hard-wired, simply do not like to interact with so many people. That's particularly the case with strangers, as this drains their energy and results in their not enjoying "sales work." In the end, they produce poor results.

AMBIVERTS

Given their innate ability to shift as needed, ambiverts, statistically, make the best salespeople. Why? Well, they don't care if you like them any more than it takes to close the sale; nor do they feel drained by the interaction with strangers.

Ambiverts are typically better listeners, too, since they don't feel a need to persuade people to like them. As a result, their communication style is more direct and to-the-point—no "show up and throw up." Indeed, the Objective Management Group has found that the top 27 percent of the world's best salespeople are ambiverts! My own 17-plus years of personal experience in behavioral science, profiling, hiring, and training nearly 1,000 salespeople brings me to the exact same conclusion.

So does, incidentally, my own personal experience. As a hard-wired extrovert who has spent the majority of my career working in some capacity of sales (salesperson, manager, VP, and business owner), I admit that I have struggled for years with the "need to be liked," taking rejection personally and "showing up and throwing up." I've had to learn to adjust my own natural way of thinking and feeling, so that I can modify my behavior to be more successful.

A word of encouragement: We're not all ambiverts. If I can do it, you can, too.

Two final factors

I'm going to conclude this chapter with two more personality factors, identifiable by the MPO, that you'll find tremendously useful as you hire, train, and retain. These are AD, for "Behavioral Adaptability" (something I was just hinting at above, if you were paying attention), and PRB, for "Perception of the Required Behavior."

Let's dive into these.

MPO AD Factor
Level of Adaptability

Provides insights into our ability to sustain change.

We can all change in some way, and often do, for finite periods.

Even the most accomplished yoga master can only hold the pose for so long.

INFLUENCES

Capacity to maintain changes in behavior.

Degree and duration of involvement.

Ability to cope with stress.

Endurance.

The AD factor:
Behavioral adaptability

I just told you how, with effort, I've been able to change my own behavior over the years. That's an anecdote. Now for the science:

Not too long ago, leaders in neuroscience discovered that we *all* have an innate ability to adapt and sustain a change in our behavior. Even better: This ability is something we can measure with an advanced personality-profiling tool. The ability to make a change in your behavior, as it turns out, requires three things:

1. **Self-awareness** that you have to make a change: "Oh, this is the task where I need to adjust my natural way of thinking, feeling, and behaving."
2. **Knowledge** of how or what to change: "I need to be more detailed, assertive, patient, etc."
3. **Motivation** to adjust and sustain a change in your behavior. This requires effort and energy, so you have to be willing to put in the effort.

As we advance and climb the ladder into management, we become responsible for a lot more things. Most jobs, after all, require us to constantly adapt or adjust our behavior, depending on the situation. Thus our ability to change and adapt is critical to our success.

That's why the AD score is one of the most important and unique aspects of the MPO, since it's one of only a few personality-profiling tools available that measures this vital trait.

The AD scale measures an individual's degree of flexibility and their ability to bounce back and adjust to various situations.

Conversely, it also reveals the degree to which people tend to focus on their business, hang in there when faced with adversity, and consistently be themselves in every situation.

The level of adaptability generally influences:

- The capacity to maintain a change in behavior
- The degree and range of involvement it takes
- The ability to cope with stress: The higher the adaptability score, the less stress a person suffers

Different positions and different work cultures will require different levels of adaptability.

A good example of behavioral adaptability

Not long ago, I was speaking to a group of CEOs, explaining to them the Three Primary Behaviors.

One of the CEOs in the audience—who had, over a 14-year period, built up a very successful company that was then doing about $12 million annually—stood up to comment.

"I can relate to this," she said. "I used to be an introvert—that is, S>E social orientation—but I'm not anymore. Today, I'm an extrovert—or 'E>S,' as you would say."

This was pretty interesting. "Why do you believe this is so?" I asked her.

"Because I'm the best salesperson the company has," she replied, rather candidly.

Given that she'd built up the company from scratch, was really good at it, and was the firm's best salesperson, I then asked her how much time, in a typical week, she devoted to selling. Her answer: About 15 to 20 percent of the time.

I then asked her what I believed to be an innocent question, especially given the "conversion to extroversion" story she had told. I asked her, quite simply: "Do you enjoy selling?"

Her face immediately darkened. "I hate it," she confessed. "It forces me to try and become somebody that I'm not. I have to paste on this happy face and pretend I'm enjoying it, when in fact, it's terribly and painfully draining."

This is what she was doing, just 15 percent of the time. "What if," I then asked her, "you had to do sales 50 percent of the time? Or what if it were your full-time job?"

She mimicked pointing a gun to her head.

"I'd kill myself," she said.

Now let me ask *you* something. Do you believe that this highly successful CEO truly *became* an extrovert? Do you think that she was able to magically morph her ingrained personality, and leave all vestiges of her innate introversion behind? Or do you think that, rather, she was simply able to "bend" herself to behave in a manner that others define as "extroverted"?

Of course she did the latter. She learned how to adjust her behavior. This underscores my earlier point: If you're born an introvert, you'll die an introvert. So for that 15 to 20 percent of her week that she needed to devote to selling, she was able to successfully—*very* successfully—adapt her behavior as needed.

This is precisely what the AD factor measures. It doesn't show *whether* an individual can alter their behavior; rather, it reveals *just how long* they can adjust their behavior as needed.

The exercise band analogy for behavioral adaptability

Remember those two other prerequisites for behavioral flexibility: 1) the *knowledge* of how to change your behavior, and 2) the *motivation* to exert the required effort. All this gets measured by the MPO's AD score.

Think of an elastic exercise band. Your synaptic connections are similar. In their resting state, relaxed, they reside at their normal length. But, like a rubber band, when stretched, they resist your efforts. They fight to return to their natural, resting state. They require "stretching" to make changes to your normal behavior. Depending on how strong (or in this case, knowledgeable) and motivated you are to put in the effort required, you will only be able to sustain the change for a period of time before your "bands" snap back to their resting state.

This is what the MPO AD factor measures: how long you can sustain the change. As I mentioned above, the very ability to detect and measure this invaluable trait is, itself, a recent scientific development, and the MPO is one of the few assessments on the market today that can capitalize on it.

By now, you should start to see, and recognize, a pattern—one that the MPO can help you deal with. How many times have you seen a manager tell an employee: "I know you can do it, because I've seen you do it dozens of times!"? How many times have you seen a manager—even yourself—ask an employee: "Why can't you do it that way all the time?"?

Now the answer should be crystal clear to you. Those poor employees may well have the knowledge, the skills, and even the motivation to do exactly what their managers are asking ("screaming"?) for. But they lack the natural hard-wired traits they need to consistently produce that desired result without "stretching," perhaps painfully, to re-orient their thinking and feeling as needed.

I guess you can see why I'm such a big fan of the MPO: It will tell you, even before you've spent any time with that individual, if and how long any given employee can sustain the required change in their personality traits to fit the required behavior of their job. This, in turn, will equip you to make better decisions when hiring, granting promotions, and coaching your employees to increase their performance and job satisfaction.

PRB: Another vital measurement

The MPO measures what it calls PRB, or Perception of Required Behavior. What do people believe the job entails? And how does that align — or not — with their natural, hard-wired personality traits and ways of behaving? This score measures that perception — and helps you to gauge that gap (or fit). It tells you how much that person will need to "stretch" to be successful in their role.

The PRB, along with the AD, will provide you with insight into the amount of effort it will take, and how long they are able to sustain the required change to be accomplish the tasks which require that change. This is critically important, because if the role requires too much effort in making these changes, the employee will a) become quickly exhausted, b) be inconsistent, c) not be happy in the role, d) eventually burn out, and e) quit, or f) you will ask them to leave. None of those outcomes is good.

But now, armed with your PRB/AD knowledge,

you can avoid this situation. You can be proactive. You can adjust the role to accommodate their needs—such as providing them with resources that will not require them to sustain those changes for as long a period of a time—and help them be more consistent and happy employees.

For example, if the person is not naturally a detail-oriented thinker, then perhaps you can provide them with a checklist or other methods to ensure that details are not missed. Perhaps you can trade some of this employee's "detail" tasks with another employee who is detail-driven and loves doing those type of tasks—and taking away those tasks that aren't a good fit with their traits. Of course, if these gaps require too much change, you may want to consider moving them to another role that is a better fit for them, rather than doing what typically happens, which is firing them!

Are you ready to hire?

You may not have considered yourself much of a scientist when you started reading this chapter. But now you have a decent-sized serving of behavioral science and neuroscience under your belt. You understand how human brains, and personalities, are wired. You've learned how that can affect job fit and performance. And you've learned, importantly, that there are tools—psychometric instruments—to help you rapidly measure and put to use these important parameters.

Let's not waste a minute! I'm going to teach you how to start applying all this science toward the key subject of "hiring." It all starts on the very next page.

CHAPTER 2

Hiring top talent

What you will learn in this chapter:

- It all starts with a structured, objective hiring process
- The candidate experience and why this is so important
- Defining success in the role
- Keeping your applicant pipeline full
- The truth about job boards
- Stop wasting time with unqualified candidates: Introducing the XP3 pre-screening process
- How to conduct objective interviews and make an informed hiring decision

- Helping the candidate make an informed decision
- Avoiding legal pitfalls
- Conducting reference checks that produce useful information

It all starts with a structured, objective hiring process

The future of your business rides on your ability to hire the right people. Every single new employee will make a difference in your company: They'll either make a positive contribution to customer satisfaction, growth, and profitability... or they'll have a negative impact on the business. You simply can't over-stress the importance of hiring great people.

But now that you know the science, you want to be scientific. You want to be objective. That's inherently difficult when it comes to, say, interviewing — the most subjective portion of any selection system. You're only human, and that becomes a disadvantage. It means you'll have to fight down your innate tendency to judge people according to how much you *like* them, in contrast to grading them according to job-relevant characteristics.

Thus the importance of a *structured* and *comprehensive* hiring process. By adopting one—that is, the *XP3 Talent* one!—you'll be taking the first step toward properly positioning your business for success by attracting the right talent. And when I say "the right talent," I'm talking about the kind of people who will help build the company culture, drive sales, turbocharge production, improve customer support, and ultimately position your company as a leader in its industry.

Sounds great, right? Well, if that's not enough of a positive motivator for you, consider the opposite scenario. *Failing* to instill the right hiring process in your company will result in underperforming employees, along with wasted resources, time, and money. A well-defined hiring process, by contrast, will streamline how you find and qualify candidates. It will simultaneously reduce your hiring costs.

Can you predict success?

Now take a look at the factors I've listed on the next page. Research has shown how these relate to your ability to make an informed hiring decision. Note, in particular, how their odds of success increase, in the order in which I've listed them:

- Education 1%[1]
- Reference checks 7%[2]
- Emotional intelligence 10%[3]
- GPA/IQ 12%
- Behavioral assessment 23%
- Structured interview 34%
- Structured interview + behavioral assessment 58%[4]

So, just how scientific can you make this? After all, you're dealing with human beings here, not machines. To paraphrase Forrest Gump's momma, "You never know what you're gonna get." Still, you need to do everything possible to make an informed, objective hiring decision.

Think of it this way: Every hire is a risk. Therefore, your job is to do everything possible to *reduce* that risk and *increase* your probability of success. And

---

[1] Hunter, J.E.: "A Casual Analysis of Cognitive Ability, Job Knowledge, Job Performance, and Supervisor, Ratings." In E Landy, S. Zedeck, & J. Cleveland (eds.), *Performance Measurement and Theory*, pp. 257-266. Hilldale, N.J.: Erlbaum, 1983.
[2] Hunter, J. & Hunter, R.: "Validity and Utility Alternative Predictors of Job Performance," *Psychological Bulletin 96*, pp. 72-98, 1984.
[3] Smith, F. & Hunter, J.: "The Validity and Utility of Selection Methods in Personnel Psychology: Practical and Theoretical Implications of 85 Years Of Research Findings," *Psychological Bulletin 124*, pp. 262-274, 1998.
[4] Smith, M. & Smith, P.: *Testing People at Work - Competencies in Psychometric Testing*. Blackwell Publishing, UK, 2005.

while no method is 100 percent effective, the XP3 hiring process, according to the above research and statistics, will provide you with an *88 percent probability of success*. That's pretty darn good!

Similarly, *all candidates* must go through the process. You need to make sure this happens.

I can't tell you how many times I've seen, unfortunately, companies skip steps in the process. Why? It could be because they already know the person. Or that candidate came highly recommended. Or perhaps they were intimidated by a seemingly-strong candidate and were scared that he or she wouldn't go through the process because they were so "special" that they could skip it.

I'm not going to mince words here. *I don't care if it's your own mother!* Everyone, and I mean *everyone*, must go through the process. All of it. It's the only way to minimize the risk of failure, and maximize the probability of success.

By the way, here's some insight into those supposedly "special" candidates I'd mentioned above: I've learned that if a candidate is not willing to go through the process, they are typically the exact same people who are also unwilling to come in to work early when needed, stay late, take on extra assignments, etc. 'Nuff said.

The candidate experience and why it's so important

Every candidate will be anxious, nervous, or both.

Think about the last time *you* felt that way. Were you at your best? Could you answer an interviewer's questions to the best of your ability? I can safely guess that your answer is "No"; that's pretty much the case for everyone.

Now think about your objective as a hirer. Isn't it to determine if the candidate has the required knowledge, skills, and behaviors to be successful in the role—and ensure that they'll be a good cultural fit for your company? I'm guessing that your answer to this question is a resounding "Yes."

Put it all together, and it means that *you need to relax the candidate so they have the opportunity to be their best for you.* Remember: A critical element of the hiring process is ensuring that the candidate makes a good decision—not just you. Is this the role and company that *they* want?

Most recruiters and hiring managers only focus on *if* the candidate can do the job. But you also need to focus on if they *want* to do the job.

To accomplish this, follow these guidelines:

- Welcome them and thank them for their time, effort, and interest in exploring the opportunity.
- Explain to them that you're genuinely interested in their making an informed decision, so you want to be sure that you answer all their questions.
- Offer them a drink or find out if they need to use the restroom before you begin.
- *Never* interview from behind your desk. That is *your* power position. Ideally, you should interview at a small round conference table. If not available, *don't* sit across from them. Seat them at the head of the table, with you and anyone else involved on the same side.
- Constantly be aware of your facial expression and body language. Be open and friendly—smile a lot!
- Outline the agenda so they know what to expect.
- Be respectful of their time. End on time.

A logical, sequential process

Here, briefly, are the steps of the XP3 hiring process:

1. Create the XP3 Position Profile, using the XP3 app to identify requirements of the role, such as knowledge, skills, behaviors, success criteria, comp/benefits, etc.
2. Create the applicant Screening Questionnaire (SQ).
3. Write a creative job post and place it on multiple job boards.
4. Pre-screen candidates: Send them the MPO and the SQ.
5. Review candidate/Prep for interview: Review the MPO, resume, and score the SQ.
6. Conduct interviews: First by video (one hour); next one in person; others as needed.
7. Conduct any other skills assessments, reference and background checks.
8. Make a verbal offer, discuss details and then send a written offer letter.
9. Create the on-boarding plan and training program.
10. Utilizing the MPO, identify communication, learning style, and motivating needs of your new employee.

Now here's what makes it all so logical and easy for you to implement: Each step of the XP3 hiring process leads to the next. Each step was designed to build on the previous one, gathering more data points as you proceed—all leading to your ability to make an informed, objective, and yes, *scientific* decision.

It's a gated process. No candidate can advance until they clear the previous hurdle:

1. The resume review determines if the candidate gets sent the SQ and the MPO. The resume review does not determine if a hiring decision is made.
2. The SQ and MPO analyses determine if that candidate advances to the video interview. Similarly, the SQ and MPO analyses do *not* determine if a hiring decision is made.
3. The video interview determines if they proceed to the in-person interview. You can figure this out already: The video interview does *not* determine if a hiring decision is made.
4. The in-person interview determines whether a follow-on interview is needed, or if a hiring decision can be made at that time.

Now that you understand how this works, let's

walk through all the key components of the XP3 hiring process.

The XP3 Position Profile

To provide you with a detailed strategy for hiring the right person for the role, we have created the XP3 Position Profile. It's a feature that's included in the XP3 app, and it helps you ensure that you gather all of the necessary data for conducting a successful search, and making an objective hiring decision.

Here's what it helps you to collect and understand:

DETERMINE THE REQUIREMENTS OF THE ROLE

It might be helpful to understand this concept by seeing what it's *not*.

A few years ago, I was approached by a VP of Sales who was frustrated with his internal recruiters. "They can't find me any good salespeople!" he complained.

And so, logically, I asked him: "What exactly are the requirements for the role, which you've given to your recruiters?"

He didn't miss a beat: "I told them I need go-getters! Salespeople who are self-motivated! Guys and gals who will do what it takes to get the job done! The kind of people who won't take 'No' for an answer!"

This happens to me a lot. I had to let him down gently. "The reason your recruiters are failing you is because you're failing them," I told him. "You're not providing them the right information to use to find you good salespeople."

He hardly seemed convinced: "But I just *told* you *exactly* what I'm looking for!"

"No," I said. "Finding qualified people for any role, begins with clearly understanding what is required to be successful in that role. What you just described to me is a list of *behavioral attributes* that *may be required* to be successful in the role, but that's not how you 'find' candidates."

You can succeed where that frustrated VP of Sales had failed. To find top talent, you need to be able define what "top talent" is. And that begins with defining the *three foundational strengths* that indicate a consistently productive and happy employee:

The 3 Foundational Strengths

that create **consistent performance**
and an **engaged happy employee**

1 **RESULTS**

2 **KNOWLEDGE**

3 **BEHAVIOR**

1. **Results.** This refers to the *application* of the specific knowledge, defined below, to a set of skills that produce consistent results. In his best-selling book, *Outliers*, author Malcom Gladwell talks about the "10,000-hour rule" that he believes is required to master any skill. That inspired me to conduct my own research, and here's what I found: It actually takes about three to five years for the average person to achieve this level of consistency in most jobs.

   Think about it. For anything that you do really well, how long did it take you to achieve that level of consistency—or "mastery," as Gladwell calls it? Now apply this knowledge to your choice of a new hire. How soon do you need for them to be producing at this skill level? Can you afford—do you have the time for—a three- to five-year learning curve? Do you have the ability to train and coach your new hire to acquire the required skills? Have you ever asked a candidate a skills-related question, only to hear: "Well, I can't do that now, but I'm smart and I can learn"? Well, that's great if you have the time and skills to teach. But if you don't know how and/or can't teach them, then you're setting yourself—and that person—up for failure if you hire or promote them.

2. **Knowledge** specific to the requirements of the role. This can include particular equipment and/or computer programs—and not what most job descriptions include ("be proficient in MS Office suite"). Most companies/employees only use three or four of the suite's seven to eight applications (*e.g.*, they'll chiefly rely on Outlook, Word, Excel, and PowerPoint), so list the ones you actually need that person to be proficient in. And be specific when it comes to the "proficiency" for each. I like to use words like "novice," "intermediate," or "expert" skills. Specific knowledge may also include knowing a specific process, such as sales, production, financial, recruiting, the XP3 Talent System, and so on.

3. **Behavior and traits.** The third key strength is the *alignment* of the required behavior of the role with that individual's intrinsic personality traits. These can include behaviors such as attention to detail, assertiveness, patience, analytical thinking, and so on. (The MPO, for example, has identified 46 potential behavioral competencies.) That candidate

can have all the knowledge and skills required to be successful in the role; however, if there isn't an alignment between the required behavior and the individual's intrinsic personality traits, you'll have problems: You won't get consistent performance. The candidate won't enjoy the role, since they'll have to work too hard to constantly adjust their own behavior "painfully stretching those exercise bands."

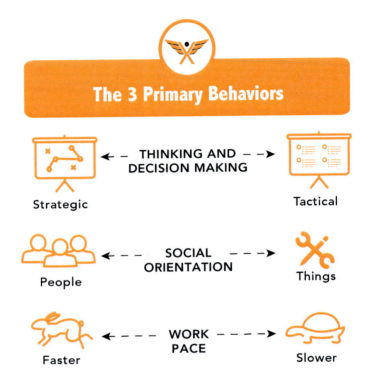

The 3 Primary Behaviors

THINKING AND DECISION MAKING

Strategic

Tactical

SOCIAL ORIENTATION

People

Things

WORK PACE

Faster

Slower

The three primary behaviors

What are the three primary behaviors that are required for the specific role you're seeking to fill? In true XP3 Talent fashion, the solution harnesses the power of "three." To identify the three primary behaviors, answer these three simple questions:

1. **Thinking and decision-making:** Does a majority of the role require more *strategic* thinking or *tactical* thinking? (In MPO terms, is it more A>S, or more S>A?)
2. **Social orientation:** Does the role require mostly working with *people* or with *things*?(Again, on the MPO side, is it more E>S, or S>E?)
3. **Work pace:** Does the role require more of a *fast, urgent pace* (A>P), or *slower, methodical, steady pace* (P>A)?

For a more detailed approach, I recommend utilizing the MPO job-profiling tool that we discussed in Chapter 1. Short of that, recognizing and leveraging the above three primary behaviors will enhance your efforts dramatically.

Defining success in the role

This is one of the single most important elements to be determined. Yet it's often the most ignored!

If you don't clearly define "success" in the role, how will you—or the employee—know if they're being successful? This is like asking them to win a race without telling them where the finish line is!

(The lack of a "success" definition is also one of the reasons why annual reviews are such a pain in the neck. The manager is busy telling the employee why they should be more successful, while the employee is busy arguing that they already are. We'll discuss this more in the Chapter 4: Performance Coaching.)

Every role hands you the opportunity to identify a variety of success measures. For example, you might be able to measure a receptionist's success by his or her ability to do things like:

- Answer the phone in fewer than three rings.
- Respond to voicemail within two hours.
- Ensure that all guests are greeted immediately upon their arrival with a friendly, welcoming response, asked if they would like a drink or use the rest room, etc.

When identifying success measures, use the SMART formula to help you be as specific as possible:

SMART Goals

S   Specific

M   Measurable

A   Achievable

R   Relevant

T   Time-Bound

Do not use vague measurements, such as "timely," "proficient," "accurate," or "satisfaction." Everyone has their own interpretations of these words, so they won't do you any good.

- For example, I see "timely reporting" used a lot. Don't go there. Instead, define specific goals like "I need the sales pipeline report by close of business every

Friday," or "I want the P&L and balance-sheet reports by the 10th of every month for the previous month."
- Don't say "Accurate." Say "100 percent" or "zero errors."
- Don't say "Satisfaction." Say "Consistently deliver a 95 percent customer-satisfaction score."

For my recruiting teams, for example, I started them off with *daily* success measures. Then, as they demonstrated they could accomplish those, we moved them on to *weekly* and then *monthly* success measures. We also tied these to their compensation, which helped them to self-manage. Some examples of these success metrics included:

- Identify a minimum of four new viable candidates each week.
- Conduct a minimum of two interviews each week.
- Fill all roles within 45 to 60 days.

Key attributes of top talent, or: How to identify an "A Player"

As I'd mentioned in the intro chapter, it's become common to lump employees into any one of three levels of talent:

- **A Players.** These are your best employees.
- **B Players.** These are the majority of your employees.
- **C Players.** These are the employees you need to immediately help find another job.

So what defines an "A Player"? You may be surprised to learn that they're not necessarily the best at what they do—although many times they are. What I believe makes someone an A Player is the fact that they are *always trying to be better*.

Here's a quick "Are they an 'A Player'?" test. First, here are some questions you can ask of the candidate:

- What do you do to ensure your knowledge and skills are at the cutting edge of your profession?
- Tell me about the last time you had to learn something new. What was it? When was it? How did you learn it? Share with me how that helped you.

Now here are some questions you can pose to yourself about the candidate:

- Have they been promoted at least once in a previous role?
- Can they speak about your company?

Can they tell you what they like, what they might change, and why?
- Are they confident without being cocky?
- What about the quality and quantity of questions they ask *you* during the interview?

Positive, detailed responses to these questions will help you decide whether or not the person is an A Player.

A Players hire A Players.
B Players hire C Players.

This is another positive aspect to an A Player that you should be aware of: A Players hire other A Players. Why? Because they realize that together, they can accomplish amazing results.

B Players, on the other hand, hire C Players. That's because they're scared of A Players making them look bad or, God forbid, even taking their jobs!

Keeping your applicant pipeline full

First, know that there are two types of candidates: active and passive.

Active candidates are those who are actively seeking a job; your job posts (advertisements) will typically work in locating these applicants.

Passive candidates, on the other hand, are those who are *not* seeking a new role; thus you must "find" them. This is true of most "A Players," but not always. Once you find them, you must then convince them to explore your opportunity. In the recruiting world, this is typically known as "sourcing" or "headhunting."

To accomplish this, you must have access to the databases of several job boards like Indeed, ZipRecruiter, Monster, etc.; this can be costly. It also typically requires a recruiting license to LinkedIn. This can be costly, too. You'll also need some expertise on how to navigate each site and do the search, a lot of time and "sales" skills to convince the potential candidate to want to "explore" your opportunity. This is typically when you will need a professional recruiter to help you.

To fill your applicant pipeline, you need to think like a marketing person. You must make your ad, and your company, stand out from all the other job advertisements or posts that typically just list the job description. Here are a few things you need to do to make your ad stand out from the masses:

Start with a creative headline. Here's an example:

JoMat Design Remodeling is seeking a results-driven, confident, energetic, and highly organized Home Remodeling Production Manager who has the experience and skills to take us to the next level of success!

Then describe, in a brief sentence or two, the role or what you are seeking.

Next, focus on WIIFM ("What's In It For Me"). People don't care about your company until they know what's in it for them!

Use language that will attract the behavioral/personality types you seek. For example, if you're looking for a High P (patient/methodical) person, keep in mind the key motivators for this score on the MPO—more on that in Chapter 4. For a High P, this includes *stability*. So if that's what you're looking for in a candidate, use your ad to show them just how stable your company is—as opposed to, say, talking about your fast growth, which could actually push them away.

On the other hand, if you're trying to attract Low P's (fast/urgent pace), then definitely tout that fast growth! Similarly, if you need to attract High A's (independent workers), then talk about the independence and autonomy the role provides. For High E's (extroverted team players), talk about the collaborative team environment.

Here's an example of a creative job post:

Residential Remodeling Production Manager

JoMat Design Remodeling is seeking a results-driven, confident, energetic, and highly organized Home Remodeling Production Manager who has the experience and skills to take us to the next level of success!

This is a permanent full-time production-management position that will be responsible for coordinating with crews and subcontractors to perform quality work. The Home Remodeling Production Manager will possess the ability to bring jobs in on-time and on-budget, and possess excellent customer relationship skills and the ability to sell additional work orders. This person must be extremely knowledgeable about, and take the lead on safety, ensuring compliance with current building codes and best practices at every job site. This person should reflect the values and dedication that we would all expect from ourselves, as if we were working on our own homes.

Some of the benefits to you:
- Competitive salary and bonus plan
- Full benefits, including medical, dental, 401k plan, paid vacation, gas reimbursement, and laptop
- Stability and growth = advancement opportunities
- Excellent training and support

What You Need to Possess:

You must have at least five years of construction/home-improvement project/production management experience, crew and subcontractor management experience, excellent computer skills including expert-level MS Word and Outlook experience, experience with job costing and change orders, a high level of integrity, creativity, and attention to detail, the ability to learn and be coached, function well in a collaborative team environment as well as work independently (self-manage), and possess an entrepreneurial spirit to roll up your sleeves and do what it takes to manage our fast growth.

If this sounds like you, then we want to hear from you today.

Our hiring process that you will go through is an example of the type of organizational process and procedures that we employ throughout our organization to ensure that we are always hiring the best of the best, while simultaneously providing you the same opportunity to make as good of a career decision as we are seeking to make.

JoMat Design Remodeling is an equal opportunity employer. We celebrate diversity and are committed to creating an inclusive environment for all employees.

All qualified applicants will receive consideration for employment without regard to the individual's race, color, sex, national origin, religion, age, disability, genetic information, status as a military veteran, or any other characteristic protected by applicable law.

The truth about job boards

Go wide. First things first: Don't put all your eggs in one basket. You need to utilize *multiple job boards* to succeed. I'm talking about sites like Indeed, ZipRecruiter, Monster, and so on. Cast a wide net, because you don't know where people are searching, given all their choices.

Leverage keywords. When people search for jobs, they'll enter keywords, such as the job title and skill set. Therefore, you must make sure your job post includes these keywords. Put them at the top of the ad. Repeat them several times within it.

Don't get buried. With each passing day, your job post will sink deeper and deeper into the depths of that site, never to be found again. Don't let that happen to you. Refresh your job post at least once a week. Change the header so it looks a little different. You need to work the site to keep your ad fresh.

Other sources to tap for talent

Post your job on your website's Careers page. ...You have a Careers page on your website, right? If you don't, create one. And remember: this is a marketing exercise. So borrow a playbook from the pros: Google top career pages and copy their style, with the aim of "attracting top talent," vs. "listing available jobs."

Another invaluable resource for finding new talent is your own employees (assuming they're engaged and happy—more on that in the next chapter). Offering a hiring bonus will motivate them to take action.

But don't do it indiscriminately. For most hiring bonuses, the employee has to wait six months to get paid; that's not terribly motivating (read: "Out of sight, out of mind"). Unless they happen to be the new-hires manager, they have no impact on how long a referred employee will stay, or how productive they will (or won't) be. So why make them wait?

Here's how to do it. Break up the bonus. Offer your referring employee a small portion of the bonus after the new hire completes their first 30 days on the job. Then pay the balance in 90 days. That's a lot closer to the "instant gratification" that most referrers seek.

You also need to communicate to your employees about open positions, and the referral bonus, frequently. Hang a poster in the break room. Send a company-wide email every time you have a new opening—with a reminder about that exciting bonus. Celebrate the bonuses paid out to employees by mentioning them in emails, the company newsletter, etc. Keep it alive and real in the minds of your workforce.

Stop wasting time with unqualified candidates: Introducing the XP3 pre-screening process

If you've ever interviewed more than a dozen candidates, I can guarantee that you've walked away from more than a few grumbling to yourself: "What a total waste of time!"

Don't let that ever happen again—or at least, minimize the odds that it will. Here to help you is the XP3 Talent System's structured *pre-screening process*. It includes tools to help you identify whether (or not) the candidates possess the requirements for the role; it will help you weed out the majority of the energy-sucking, time-wasting candidates, ensuring that you're only spending time with the ones who matter.

The XP3 scientific and objective pre-screening process

1. Review resumes against the requirements of the XP3 Job Profile.

2. Conduct screening calls with candidates.
3. Administer an XP3 knowledge and skills Screening Questionnaire (SQ) and MPO survey.
4. Analyze the MPO against the MPO job profile to identify fits and gaps.
5. Review the resume in more detail to identify relevant experience and education
6. Score the XP3 Screening Questionnaire.
7. Schedule and conduct a one-hour video interview.

Now let's walk through the details of each of the above steps.

Reviewing resumes

The first thing you need to understand about resumes is this: *You don't hire resumes. You hire* **people**.

Keep that resume in its proper perspective. It's merely a first step in determining who you will advance to the next step of the hiring process.

Some of the best people/top talent, incidentally, lack great resumes. Why? Because they don't need one. Here's a story from that realm, one that's close to home for me:

In her entire career, my wife has only had four jobs. Four. She had to find the first one. But after that, her boss took her with him to her next two jobs. Then he retired. So, after all these years, it was finally time for my wife to look for a job again—only the second time in her entire career. Just before she started applying, I asked her if she'd like me to review her resume for her.

She agreed, albeit reluctantly. "What's so important about a 'well-structured resume'?" she asked me. "I never needed one before."

"Well you sure need one now!" I blurted out after looking hers over. "Do *not* send this out to anyone!"

She was surprised. "Why not? It shows my education, all my job titles, and bullet points of information for each of those jobs. What's so bad about that?"

"It's not what's there," I explained. "It's what's not there. Where are those killer accomplishments you've achieved over the course of your career? Where are the career progressions, the promotions in each role? Where are the keywords that recruiters depend upon to help them find top talent?"

Needless to say, she ended up with a far better resume.

But my wife's story represents just one end of the spectrum. I'll bet you're familiar with the other end. Haven't you seen some beautifully-written and perfectly-structured resumes—only to interview the people who "wrote" them and sense a distinct disconnect? Of course you have. These days, anyone can hire somebody to whip up a resume for them that makes them look like a rock star.

Based on that "spectrum" I've described above, here's how I suggest you proceed:

Reviewing a resume in three easy steps

1. Scan for the keywords that match the job requirement's "must-have" knowledge and skill sets.
2. If applicable, look for the educational requirements.
3. Look for a match of the required years of experience needed.

Key Words

Jane Doe

1234 Street Address, City, ST. Zip Code ~ 700.000.1234 ~ janedoe@address.com

Executive Summary

A seasoned and passionate professional with 30 years of experience and expertise building and leading top performing teams. Expertise includes relationship management, operational efficiencies, process improvements, marketing, risk analysis, financial management, project management, regulatory & HR compliance, internal auditing and de novo bank start-up. I have built a reputation for being a fast-paced, proactive, results-oriented leader who is known for superior customer service and complex problem resolution.

Objective

Years on Job

Jane Doe

1234 Street Address, City, ST. Zip Code ~ 700.000.1234 ~ janedoe@address.com

Executive Summary

A seasoned and passionate professional with 30 years of experience and expertise building and leading top performing teams. Expertise includes relationship management, operational efficiencies, process improvements, marketing, risk analysis, financial management, project management, regulatory & HR compliance, internal auditing and de novo bank start-up. I have built a reputation for being a fast-paced, proactive, results-oriented leader who is known for superior customer service and complex problem resolution.

Objective

Education

Education

American Institute of Banking – June 1989 – graduated with honors
Virginia Highlands Community College – September 1984 to September 1986

Licenses Obtained
Series 6 & 63

That's it. If two of the three are a mismatch, then move on to the next resume.

If two of the three are a match, determine if the one that is *not* a match can be overlooked due to an abundance of experience and stated accomplishments. Then move to the next step in the pre-screening process. If not, then move on to the next resume.

If all three are a match, move to the next step: The screening call. You will review the resume in more depth prior to the interview (which we'll discuss in the interview prep section, shortly).

Candidate screening calls

Now that you've built your pile of "Potential" applicant resumes, it's time to put on your Sales hat.

Why? Because that's what the next step in the process is all about.

You'd be amazed at how many people miss that crucial point. Most of the initial applicant calls that I hear hiring managers make, make me cringe. They come across as if the *manager* is doing the *applicant a favor* by calling them! It's all one-sided: "Here's what we need. Here's what we want." Uggh.

Do what I do. Flip the tables. *Assume that every person you'll speak with is a killer A Player.* And since we'll be utilizing a structured hiring process that every candidate must complete, you'll need to get that "A Player" excited about the opportunity that exists at your company. You want to get them so amped up that they'll be willing to jump through hoops to get a job with you.

So Lesson One is: *Never ask them if they're interested in a job.* Most A Players already *have* a job, and/or are *not* seeking new career opportunities. Remember: You're wearing your Sales hat. Similar to your job posting, you need to focus on WIIFM ("What's in it for me") for the applicant. Here's an example of an XP3 phone script:

THE XP3 PHONE SCRIPT – A SAMPLE

Hello [First Name],

The reason I'm calling is because we received your resume in consideration of the Project Manager opportunity at JoMat Company, and it appears that you may be an ideal fit for what we are seeking.

JoMat is an award-winning, 30-year-old company that's ranked in the top five in our industry. We are growing—and so we need someone with your skills in order to help us better achieve our growth objectives. We offer extensive training, above-average compensation, a generous benefits package, and opportunities for advancement. Does this sound like something you'd be interested in exploring?

As you can hopefully appreciate, we're only seeking "A Players." To that end, we've developed a fairly extensive hiring process to ensure that we're truly identifying the best of the best. But that also means that we're just as focused on helping you make an informed decision. We want to ensure that we're a good fit for you.

To help us both move forward in our hiring process, we'd appreciate your completing a couple of questionnaires. The first one is a knowledge-and-skills questionnaire; the other is a brief personality profile called the MPO.

It will take you about 30 to 45 minutes to complete both questionnaires. Once we receive your results, we'll let you know if you will pro-ceed to the first-round interview. So how soon do you think you could get these completed?

Great! I'll send the questionnaires your way. Be sure to read and follow all the instructions carefully. Also be sure to check your grammar, as it will count just as much as your answers.

Do you have any questions that I can answer for you?

Thanks, [First Name]. We appreciate your time and your interest in JoMat. We look forward to reviewing your results, and hope to meet you in an in-person interview soon.

The XP3 applicant Screening Questionnaire (SQ)

The XP3 applicant Screening Questionnaire, or SQ, is a powerful tool. It was designed to identify whether (or not) applicants possess the required knowledge and skills to be successful in a given role. It helps identify a handful of behavioral attributes, and signs of intelligence that lead to success in the role.

The SQ is also continuously being improved. Over the last decade, we have assembled a library of more than 100 role-specific questionnaires; they're now included as part of the XP3 Talent System web-based app.

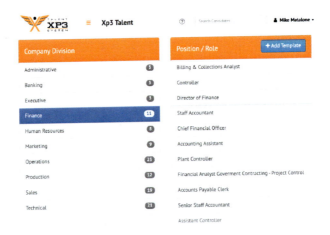

The SQ is sent out after the screening call. I'd mentioned that it screens for required knowledge and skills for the role, as well as "signs of intelligence," and the ability to learn and be coached. But how does it do all that?

It's as clever as it is straightforward. First of all, it measures the applicant's ability to simply follow the instructions. Do they exhibit attention to detail? If they provide specific, detailed responses, and check their grammar, the answer is "Yes."

The SQ also helps you to identify some critical "A Player" attributes, including a sense of urgency and the willingness to put in a little effort to win the job. The expectation is that a candidate would put forth their best effort in order to obtain the job; this is typically the benchmark for what you can expect of them once they're actually on the job.

Hire • Train • Retain

XP3 Screening Questionnaire Example

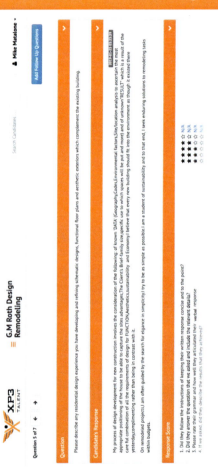

XP3 TALENT ← ↑

Question 5 of 7 ← ↑

≡ **G.M Roth Design Remodeling**

Search Candidates...

▲ Mike Matalone ▾

Add Follow Up Questions

Question

Please describe any residential design experience you have developing and refining schematic designs, functional floor plans and aesthetic exteriors which complement the existing building.

Candidate's Response

My process of design development for new construction involves the consideration of the following: of known 'DATA' (Geography,Codes,Environmental factors,Site/location analysis to ascertain the most appropriate positioning of the house to be able to capture the sites advantages;The Client's Brief-family size,specific use to which spaces will be put and more) and of unknown"RESULT" which is a result of the care ful combination of all the requirements of design for FUNCTION,Aesthetics,sustainability and Economy.I believe that every new building should fit into the environment as though it existed there yesterday,complimenting rather than being in contrast with it.

On remodeled projects,I am often guided by the search for elgance in simplicity.I try to be as simple as possible.I am a student of sustainability and to that end, I seek enduring solutions to remodeling tasks within budgets.

Response Score

1. Did they follow the instructions of keeping their written response concise and to the point? ★★★★☆ N/A
2. Did they answer the question that we asked and include the relevant details? ★★★★☆ N/A
3. Please rate their grammar and how well they articulated their verbal response ★★★☆☆ N/A
4. If we asked, did they describe the results that they achieved? ☆☆☆☆☆ N/A

78

The XP3 Screening Questionnaire video question

Another great feature of the XP3 Screening Questionnaire is the video screening question. This allows you to view a brief, two-minute video of the candidate explaining why you should select him or her. While their written response to the SQ helps you gauge their ability to communicate effectively in writing, the video question helps you to see—and hear—if they can communicate effectively in the verbal realm.

XP3 Video Question

9. Video Question

Please answer this question using your laptop, ipad or Smartphone camera where you will have up to 2 minutes to tell us *why we should select you for this role and what do you attribute your sales management success to?*

How would you like to answer: ?

What about candidates who are unwilling or unable to complete the SQ?

That's a great question. But the answer is self-evident. In my experience, if the person is not willing to devote 30 to 45 minutes of their time to answer a few questions and possibly land a new career opportunity, then that's the *exact same person* who won't come in early, stay late, or take on additional projects.

In fact, the SQ is purposely designed to weed out those types of candidates. It's also designed to save you from dealing with candidates who agree to complete it, and then *can't*, because they lack the knowledge and skill required to answer the questions—and thus be successful in the role.

So with all this weeding, trimming, and culling, you can expect about a 70-percent completion rate. That is, for every ten you send out, you'll get about seven people who actually complete it. The other three are already out of your hair.

The SQ is more than a pre-screening tool

The SQ isn't just for pre-screening. It's used during interviews, too.

I've seen lots of completed SQs that simply *glowed*: worthy of a five-star review. But that can be problematic. How do you know if the applicant actually wrote those responses? How do you know they didn't Google the answer, or call a friend for advice?

Years ago, I received an SQ from an aspiring sales manager whose answers were of the "five-star" variety. Reading his responses, I kept thinking, "Wow, that's exactly how *I* would have answered those exact same questions!"

And then it hit me: *I had! Those were my answers! Verbatim!*

But, being the optimist that I am, I assumed that this candidate had probably heard me speak, or had read something I had written. That's when I realized where he'd gotten his answers. I'd had a white paper on sales management posted on my website, and this candidate had simply copied-and-pasted *my* text into *his* SQ.

I still clung to my optimism. "Well," I reasoned, "that was fairly creative of him—and 'creativity' isn't a bad attribute for a sales manager." So I decided that I'd simply interview him and ask him the exact same questions that he'd "answered" so brilliantly in the written SQ. Could he do it live, impromptu, with the same easy grace?

No. He could not.

With my optimism admittedly dinged, I realized that *this needed to become part of the XP3 process*. Think about it: If a candidate can provide good written responses to the SQ, and then back that up with equally-good verbal responses to *the exact same questions*, then that candidate clearly possesses the knowledge and skill required to be successful in the role.

Therefore, to this day, a critical component of the XP3 hiring process is to ask the candidate at least three of the most important knowledge- and skills-based questions from the written SQ, during the interview, to confirm that he or she actually knows what they'd submitted to you in writing. Remember: Our objective here is to obtain as many data points as possible that steer you toward a successful hiring decision. You're not looking for photographic memory, or rote recital of what they wrote down, word-for-word. You simply want to ascertain that the candidate knows the answers to the questions that pave the way to success.

Scoring the SQ

How, then, do you score the SQ? You use the XP3 app. (Your purchase of this book includes a 60-day free trial of the web-based app.)

The app comes complete with an objective, analytical scoring system. Each of the candidate's responses is rated on a five-point scale—which ranges from Poor, to Fair, to Good, to Very Good, to Excellent—and results in an overall numerical score. This helps ensure that you can make an objective, informed decision when it comes to the next hurdle in the selection process: Whether (or not) you should advance that candidate to a video interview.

Using the web-based app, each hir scores the candidate's responses, using t scale described above. The manager's scor into the system, along with an aggregate av to help you conduct in-depth candidate analyses. As a guideline, a viable candidate should have an average score of 3.0 or higher, in order to qualify for the next step in the process.

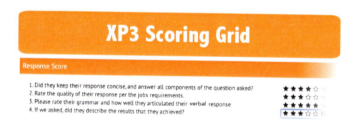

XP3 Scoring Grid

Response Score

1. Did they keep their response concise, and answer all components of the question asked?
2. Rate the quality of their response per the jobs requirements.
3. Please rate their grammar and how well they articulated their verbal response
4. If we asked, did they describe the results that they achieved?

XP3 SQ
Candidate Analytics

Candidate Comparison   **Hiring Manager Comparison**   By Hiring Manager

Below you can compare how different hiring manager have evaluated the same candidate.

Select Candidate

Sarah Connors

Action	Last Name	First Name	SQ	Interview	Final	Notes
👤	Matalone	Mike	4.2414	4.8182	4.4000	NOTES
👤	Bates	Mary	3.8065	2.8333	3.5349	NOTES
👤	DeWulf	Leigh				

XP3 Applicant Tracking

Candidates for Vice President of Sales   👤 Add Candidate   ⬇ Export

Name	SQ Sent	SQ Answered	SQ Score	Interview Score	Avail Start Date	▲ Status
👤 Duke, Donna	04/08/16	04/09/16	3.0690		5/1/16	Follow-up Interview Scheduled
👤 Matalone, Michael	09/03/19	02/13/19	4.8077		2/27/19	Hired
👤 O'leary, Josuah	04/08/16					New Candidate
👤 Sellers, Joe	04/08/16					Sent SQ & MFO
👤 Vooder, Donna	03/07/18	03/07/18			3/13/18	Stopped/SQ/MFO
👤 Connors, Sarah	02/15/19	03/07/16	4.0167	3.7826	3/30/16	Video Interview Scheduled
👤 User, Sam	02/27/18	02/27/18			2/26/18	
						Withdraw/Not Responsive

The XP3 Response Guideline

So far, we've talked about the actual scores that you and your hiring managers may assign to a given candidate's responses. But how do you know what makes one response "Good," vs., say, "Excellent"?

Again, the XP3 Talent system is here to help you. Many of the screening questions come with a *Response Guideline*. This valuable resource helps anyone who's tasked with scoring applicants' responses to make objective, consistent judgments. In short, the Response Guideline lets you know what to look for—the elements that a good answer should contain.

This, incidentally, lets almost anyone be a competent "scorer," even if they lack the subject-matter expertise being covered in the SQ questions!

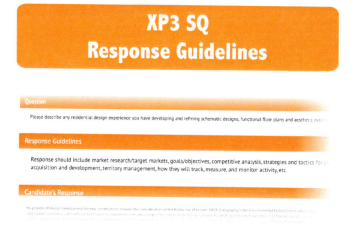

XP3 SQ
Response Guidelines

Question

Please describe any residential design experience you have developing and refining schematic designs, functional floor plans and aesthetic exte...

Response Guidelines

Response should include market research/target markets, goals/objectives, competitive analysis, strategies and tactics for p... acquisition and development, territory management, how they will track, measure, and monitor activity, etc.

Candidate's Response

My process of design development for new construction involves the consideration of the following of known 'DATA' (Geography), index, Environmental factors, Site location... applicable parameters of the known hardscape because the client's advantage. The client's flight for an out specific use to which client will benefit and the reac...

Analyzing the MPO for job fit

You'll like this part: It's where the science comes into play. As we'd discussed in Chapter 1, the MPO suite of tools includes a Job Profile, which helps you establish the ideal and/or required behaviors for a given role. This becomes your benchmark. You'll use it to gauge the suitability of the individual's MPO profile. The MPO even includes a tool, called Right Match, that does the analysis for you.

That said, it's important to remember that *the MPO does not determine if a person can do the job.* Rather, it can reveal a *drive to behave in a particular way*—a person's intrinsic motivations and therefore their desire to do the job.

Review the resume in more detail

Now review the resume for the following:

- Look at their education level and take that into consideration regarding how the resume is formatted, and how they responded to the SQ. For example, if they have a college degree, you should expect stellar grammar.
- Look at the time on each job. Short tenure does not necessarily mean they're a flake, but you do need to ask why and look for patterns. Also consider the reasons they say they left. Ask yourself if you're putting them into a similar situation—if so, don't hire them.
- Identify anything that you want to know more about during the interview.

Summary of the XP3 pre-screening analysis

By now, we've come full circle. We began this chapter with a review of the three foundational strengths that create a consistent, productive, and happy employee:

1. **Results:** The *application* of the specific knowledge to a set of skills that produce consistent results.
2. **Knowledge:** specific to requirements of the role.
3. **Behaviors:** Alignment of the required behavior of the role with that individual's intrinsic personality traits.

Now, when you employ the XP3 pre-screening process, you'll be equipped with the analytical data you need to determine if—and to what degree—a given candidate possesses the Three Strengths. If they do, they'll advance toward the next step (which I'll discuss below).

Importantly, the XP3 pre-screening process also tells you when to hit the brakes. It will help you decide when *not* to advance a candidate to the next step.

But this brings up an important distinction that I

want you to remember: While the system *can* tell you to "not advance" a candidate at this point, it *can't* tell you to simply "hire" a candidate at this point. Right now, you only have enough information to make a go/no-go decision regarding the next "gate" in the system: The one-hour video interview.

The one-hour video interview

This is the next step in the XP3 hiring process—for those candidates who have cleared the previous hurdles we just discussed.

The one-hour video interview can harness any available popular conferencing technology: Zoom, GoToMeeting, Skype, whatever. Among the many things which this accomplishes is its revelation of your applicant's comfort level (or lack thereof) with some of the basic technology which is now a part of everyday business.

And just like the previous steps, the one-hour video interview is yet another hurdle for that candidate. The interview will help you make a go-for-ward/not-go-forward decision, but, just like the prior steps, it will *not* give you enough information to make a hiring decision. Got it?

I recommend that, whenever possible, you should have two hiring managers conduct this interview with the candidate. At XP3, we do this along with our clients manager. Setting it up this way allows you to have two people asking questions—and taking notes.

Immediately after the interview, conduct a debrief with your co-interviewer. Compare notes. This will help you both make an objective decision as to whether or not that applicant should advance.

Now that you know the basics about the one-hour video interview, let's get into the details.

Interview prep

Your first step is to review the candidate's MPO *personality* profile, and compare it to the MPO *job* profile. This will help you to keep their intrinsic behaviors in mind as you review everything else you need for confirmation. For example:

- If the candidate has a high S score—indicating traits such as attention to detail, organized thinking, compliance, and a hard-wired need to do things correctly— then look for those things as you review their resume, their SQ response, and how they communicate.
- If, on the other hand, the candidate has a low S score—the opposite of what I just

described—then you want to see if any of those have become learned behaviors.

This comparison of their personality traits to the MPO job profile you'd completed (which provides a benchmark for the role's required behaviors) will prove invaluable. It will help you find fits as well as gaps. It will let you identify their most hard-wired/predictable traits. And it will enable you to focus on their three primary traits, according to the scales we'd discussed in Chapter 1:

- **A/S:** Thinking and decision-making style
- **E/S:** Introvert (cautious and slow to trust, direct, factual communication) or extrovert (quick to trust, friendly, social communication style)
- **A/P:** Fast/urgent, vs. slower/methodical pace

With this "behavioral profile" in mind, now review their resume and their responses to the XP3 Screening Questionnaire (SQ). See if these behavioral traits show up in the way they wrote their resume, and how they responded to the SQ. And during the interview, see if these same traits surface when the candidate responds to your interview questions. Listen for verbal cues; watch for body language.

Special info for interviewing introverts

If the candidate you're about to interview is an introvert (S>E), you'll need to bear in mind that they're intrinsically cautious, analytical, and slow to trust. This means you'll have to work even harder to get them to open up and speak freely; see the candidate experience we previously covered.

Don't expect fast responses to your questions— and that's a good thing! Introverts will take their time before answering; and when they do, their responses will be direct and to-the-point. Don't confuse these types of replies with a lack of knowledge; you'll just need to probe for more specifics.

And remember: If "cautious" and "analytical" are some of the required behaviors of the role, then what I've just described would be very encouraging, in terms of what you're seeking.

The one-hour video interview agenda

First things first: Be sure to let the candidate know the objective of this interview. It's not about making a hiring decision. Rather, it's just an opportunity for them to get to know you, and vice versa. It's the next step in the hiring process. Therefore, you don't need to ask everything you want to know—rather, just enough to determine if an in-person interview is warranted or not.

In the agenda below, I've used "We" as the interviewers, and "You" as the candidate. This will let you copy this agenda for your own use. The parts I've indicated in brackets, [like this], are for your use only, so you'll want to delete those if you plan to share the agenda with the candidate. They only need a brief statement on each.

1. **Introductions**. We will introduce everyone who is attending the interview.
2. **Overview.** The Hiring Manager will provide a quick summary of the company and position.
3. **Any questions?** At this point, *you'll* get to ask *us* questions about the company/position, to help you make an informed decision.
4. **Cultural add (as in "add value" to the existing culture vs. "fit" into a mold).** Here's where we'll ask you to share a little about who you are as a person—the kinds of things that aren't on your resume, but will help us get to know you better.

5. **Your resume and background.** Here, we'd like to ask you about the different things you've done and achieved in your career. [Most importantly, this is where you'll learn why they left their last role, and what motivated them to take the next one. Look for patterns.]

6. **Knowledge/skills.** We will next ask you questions from the screening question-naire we'd previously sent you. [Tell them they did a great job on this, which is why you are interviewing them. Now that you are face-to-face (albeit virtual-ly), you are going to ask them some of the exact same questions that were on the SQ, with the opportunity to see them talk you through their answers. Watch their reactions; keep in mind the story I'd told about the applicant who had copied text from my white paper! Watch to see their reaction. Do they get visibly nervous or remain calm and confident?]

7. **Any other questions?** At this point, we'd love to learn if there's anything else about you that we should know, or you'd like to know about us, that we hadn't asked.

8. **How's it feel now?** Now that we've spent a fair amount of time talking and getting to know one another, do you think this is

an opportunity for you? And if so, why? [Also ask them what their three primary decision factors are for accepting their next career.]

9. **Wrapping up/next steps.** [To conclude the interview, thank the candidate for their time. Ask them to please take a day to think about it to make sure that this is something they truly want to do; and instruct them to please send you an email explaining why they want the job and why you should select them. Then let them know that you will get back to them within a few days to tell them whether or not they've advanced to the next stage in your process: the in-person interview.]

The post-video debrief

As soon as the interview concludes, and everything is still fresh in your mind, huddle with your other hiring manager(s) and discuss what went well, and what didn't. Make notes about your comments that you could clearly reference later. Once the candidate sends you their follow-up email as they were asked to do—which is another "test" about how compliant they are with following instructions—then you can then decide whether that candidate should advance to the in-person interview or not.

The in-person interview(s)

Congratulations—to you, and the candidate! If you've successfully gotten to this point, you've reached the third official company/candidate interface of the XP3 hiring process. First was the screening call. Next was the video interview. Now comes the in-person interview.

The video interview was limited to an hour. The in-person one isn't. You'll take as much time as is needed to make an informed, objective hiring decision after this interview. That said, you need to respect the candidate's time. Determine, in advance, who else you will have interviewing the candidate, and how much time *they* will need, and schedule it all accordingly, so that the candidate can plan his or her time as well. And don't forget: Anyone else on your team who is going to be part of the interview process, must also be trained and aligned on the *entire* XP3 hiring process.

While *your* goal of the interview is to *hire* the best candidate, keep in mind that the *candidate's* goal is to "be" the best candidate! That means that they're going to try to say, act, and be everything that they think you want them to be.

Have you ever hired someone, only to wonder, a few months into the job, whatever happened to "that person" you hired? It's not uncommon. It's because interviewing is a lot like dating. We're all on our very best behavior: trying to be and say whatever it takes to get the job. Then, after we get the job (or the relationship), we relax and revert back to our "normal" selves.

The XP3 hiring process is designed to help you avoid this. It's been structured to help the candidate relax through each step, so that, hopefully, at this point, the candidate is sufficiently comfortable to show you the "real person inside," and not the "representative they sent to get the job." In case they're not relaxed and forthright enough, bring them back again for another round.

Tips for tip-top interviewing

Don't dive into the "interview" part of the interview. Refer back to the Candidate Experience section for tips on how to relax the candidate and ensure they have a good experience.

Next, take the candidate on a tour of the business and introduce them to people. *Then* you can begin the interview.

As a reminder about what I mentioned during the candidate experience, *never* interview from behind your desk; that's a position of power and intimidation. Instead, interview at a table—and even then, not across from them, but next to them. Try and limit the interviewing team to just one or two hiring managers at a time. You should all sit on the same side of the candidate, so he or she needn't look back and forth.

Pay attention to your facial expressions and body language. Smile a lot—it goes a long way toward helping the candidate relax.

The outline for the in-person interview

Okay, are you ready to learn how to conduct the in-person interview? Then grab a pencil, get ready to take a lot of notes, and memorize a ton of really difficult new procedures.

Just kidding! I'm happy to tell you—and you'll be happy to learn—that *the in-person agenda is the exact same one you'd used for the one-hour video interview*. It's all part of the XP3 Talent System's design to be as easy-to-do as we could make it.

Simply follow the same process I'd described for the video interview. The only difference here is that you'll have more time. Probe deeper on any areas of concern that you perceived from the video interview.

Then, once you're feeling that this is indeed a viable candidate, it's time to switch modes and...

Help the candidate make a good decision

This isn't just about you. Or your company. One of the biggest mistakes that gets made, every day, in the world of hiring is that employers only focus on what *they* need, and whether or not the candidate can *do the job*. But if you want to make a fully-informed hiring decision, you must also determine whether or not the candidate *wants to do the job!*

Those are two vastly different things.

Think about it. There are lots of things we can all do very well—but that certainly doesn't mean that we *enjoy* doing them. Typically, if we only have to do these "un-enjoyable" tasks about 15 to 20 percent of the time on the job, then it's no big deal. However, if you have to do them, say, *40 or 50 percent* of the time, things change. Radically. Or what if you, as a candidate, learned that you wouldn't be given sufficient resources (tools, people, funds, etc.) to do the job as you've been accustomed to performing it? Would you take that job?

Let me give you a personal example. If I, Mike Matalone, were offered a role that put me in charge of hiring, training, and retaining, I would only accept it if it allowed me to use the XP3 Talent System. The reason should be pretty clear: Aside from my personal bias, it's the method that I know how to employ successfully.

Could I succeed using another system? Maybe. But I wouldn't know until I actually tried it—and that's just too risky, in my estimation. This is why, toward the end of the in-person interview, you need to switch gears and help that candidate to make a truly informed decision. Spend time to make sure that they're fully aware of what the job entails—and what your expectations are. Let them know what resources they'll get (and which ones they won't), which people they'll be working with (and their personalities and behaviors—warts and all). You want full disclosure. No surprises!

Additional interview questions

Use the following, conveniently arranged by category, to help you as needed.

CULTURAL ADD

Finding candidates who can add value to your existing culture means digging deep to uncover their core personalities. You want to learn what drives them, what their ideal work environment is like, and what passions—both inside and outside of the workplace—drive them forward. Here are some thought-starters:

- **Why are you leaving your current company/why did you leave your last job?** This question measures potential temperament and response to conflict better than any other. Oftentimes, employees leave a company because they are unhappy with something; other times, they're simply pushed out. This question gives them the opportunity to display both honesty and professionalism.
- **Tell me about yourself.** This seemingly simple request can be profoundly revealing. Will they talk about their personal life? Will they ramble? Will they freeze up? Act awkwardly? It's so open-ended, that it automatically gives you a great grasp of how they communicate under a reasonable amount of stress.
- **Why do you want to work here?** This question accomplishes a lot, at once. It

determines if the person did their home-work. It lets you know if they're passion-ate about your industry and/or company. And it shows you whether (or not) they can learn, listen, and be flexible. With one quick question, you'll find out if their core values are a good fit with those of *your* organization.

- **How do you define success?** Be genuine-ly interested in their response to this one: it often reveals a great deal about their character. Pay special attention to their initial reaction: You can usually tell, by the candidate's body language, if they've ever thought about their own personal definition of success before—or if this is the first time it's ever crossed their mind.

- **What's the most recent book you've read, and why?** Use this question to gain insights into their core personality: Are they a learner? Do they love history and details—or romance and fun?

- **What internal factors would drive you to leave a good job?** Here, you're prob-ing to uncover what specific situation(s) would derail an otherwise happy em-ployee. Is it a micromanaging boss? Lack of work/life balance? Insufficient pro-fessional development? This question will help determine their preference in working environments—and if they are potentially a fit, culture-wise.

JOB FIT

As I'd mentioned above, knowing if someone is the right fit for a job, and knowing if that job is the right fit for them, are two different things. Even if that candidate has qualifications that align with the role, the deeper intricacies of the job may not be a good fit for them. Use these questions to find out:

- **How did you prepare for this interview?** Their answer to this question will show you how interested they are in the position and the company. It will be pretty obvious if they've only taken a perfunctory glance at your website, compared to a thorough deep-dive. It also indirectly provides insight into the applicant's soft skills, like attention to detail and motivation. These are qualities that are often challenging to evaluate during an interview.
- **Explain a new idea to me.** Say something like this: "Take a complex process, product, or service in your current job, and explain it to me so well that I could teach a class on it tomorrow." Candidates

always claim to have some version of "excellent communication skills," "good oral and written communication skills," or "great people skills." This question tests those claims. As applicants reply, pose questions at various stages to see how they react. Their communication during this "explanation" will likely be similar to the way they'd interact with co-workers or customers.

- **I'm not sure if you're the perfect fit for this role.** Put another way: "Can you tell me why you think you'd be a great candidate?" This is a great question because it's an *objection*. Their response will show you how they handle objections such as those they might receive from a potential client or a co-worker with a contrarian opinion. It's also a good way to see if they've done their homework on the company, and their depth of understanding what it would take to be successful in the role.

- **What prompted you to apply for this job?** Or: "What interested you the most about this specific position?" This question will reveal the amount of research the candidate has conducted on the company before applying for the job. It also allows the interviewee to list some of his/her strengths that they believe fit well with the position.

- **What are the things that are most important to you in the next role you find yourself in?** This question isn't just about what the company needs. It's about what motivates *them*, and where they really want to see themselves next. You'll want to make sure that the technical fit is there—but also the fit for how they are planning their career and personal life.

TEAMWORK

Nobody knows how your team works as well as you do. And while some employees prefer a little more autonomy than others, most share similar values when it comes to teamwork. Therefore, if you hire someone who *doesn't* connect at that same level, you can expect a whole new set of issues for the company.

- **How do you rely on other members of your team or network to make you better?** This is a combo question. It shows how a person behaves with their co-workers (cultural fit). And it also gives you insight into the way they view their own strengths and weaknesses—without your having to actually ask, "What are your strengths and weaknesses?"

THOUGHT/WORK PROCESSES

- **What's the best piece of feedback you've ever received from a supervisor or manager on your performance?** And: "How did you apply that feedback?" This type of question gives the candidate an opportunity to show that he/she is willing to accept feedback and has the ability to utilize the information. The best answers demonstrate that the candidate accepted constructive critique—and recognized an opportunity to improve performance. It will also reveal humility and respect for managers/mentors.

- **What are the biggest ways people waste time on the job? What do you think are the reasons for this?** What you're looking for here is defensiveness, discomfort, or dishonesty in your applicant's answers. Do they cite improper workflow, lack of training, indecisiveness, non-functional teams/conflict, or insufficient equipment? The level of their analysis will clue you into their level of thinking.

GENERAL QUESTIONS

- **Other than your family, what are you passionate about?** Whether it's their Sunday night bowling league or their son's Boy Scout troop, they need to be passionate about *something*. Put another way: If the candidate isn't passionate about anything, then he or she surely won't be passionate about working for your company.

IDENTIFICATION OF AN "A PLAYER"

- **What do you do to to ensure that your knowledge and skills are at the cutting edge of your profession?** A Players are always looking for ways to up their game. They're avid learners who are driven to improve their knowledge and skills. Thus, a response indicative of an A Player can include things like reading work-related books, articles, blogs, recent training attended, hiring of a coach, etc.

RETENTION QUESTION

- **Who were your best and worst bosses — and why for each?** Remember that people don't quit jobs; they quit managers. So if they end up describing *your* management style as "their worst boss," this probably isn't going to work out. Conversely, the way the describe "their best boss" reveals how they *want* to be managed. You need to decide if you can and are willing to manage them the way they want to be managed, which better ensures you of having a happy and productive employee.

Avoiding legal pitfalls

First things first: Stick to the job requirements. The best way to avoid any legal pitfalls is to only ask questions that pertain to the goals and measurement standards of the required knowledge, skills, and behaviors stipulated in the job description.

Keep it friendly, but professional. Avoid personal questions; these include anything that have to do with marriage, family, sexual orientation, gender, religion, race, children, etc.

Conducting reference checks that produce useful information

In my experience, most reference checks are a waste of time. Calling HR provides virtually zero actionable information. Calling colleges and even a former supervisor will yield very little value, given time it takes to track them down and call them, utilizing the approach that most of us were taught:

"I was calling so you could tell me about [Name]?"

You can guess what happens next. The supervisor will automatically tell you what a wonderful person that candidate is, that they would hire them again if they could, *blah, blah, blah.*

If you really want to get the biggest bang for the buck of your time investment, try using the XP3 Reference Check method. First, you want to get references from two previous and recent supervisors: No friends, colleagues, etc. You want their immediate supervisor. Then use this script:

Hello _____,

My name is (your name) and the reason I'm calling is because (candidate's name) has you listed as a reference. I know that you're probably very busy, so I'll be brief. Is now a good time? Great, thank you for your time and feedback. To keep this short, I want to let you know that I am not calling so you can tell me how great (candidate's name) is, since we have already identified many of his/her good qualities, which is why we're considering having him/her join our team.

Instead, the reason for my call is to gain some insight on how best to train, manage, and work most effectively with (candidate's name). To that end, I have a few questions I'd like to ask you; would that be okay?

Could you please share with me what you found are the best methods for (candidate's name) to learn new things as well as any areas I should focus on?

What's the best way to get the most productivity out of (candidate's name)?

What can you tell me about what motivates (candidate's name)?

If, as they respond to your questions, you find that they're not answering them adequately, it's time for you to push and probe. Keep asking them: *"Could you please share a little more about that?"*

You'll be amazed at how much most people will share that is actually useful information. So take lots of notes. Learn from what you hear. Pay attention to how you can utilize this to better train, motivate, and get your new hire productive as quickly as possible.

Congratulations
—you're a hiring pro!

As you progress through this book—and the XP3 Talent System—you just keep getting better and better. A nice pat on the back is in order: You've now nailed the science of talent management (Chapter 1), and all the subtle nuances that add up to successful hiring (this chapter). You now know that effective hiring depends upon having a structured, objective process. And then you learned what that process is, and how to put it into action. You've learned that the XP3 hiring process includes lots of "gates" or "hurdles" to help you (and your candidates) know, confidently, whether those applicants are ready to advance to the next step in the process. You've even learned how to conduct video interviews, and leverage this knowledge for the in-person experience.

What all this means is, you're now at the point where you are ready to hire a true A Player. But do you want them to *leave?* Of course not. The next part of the XP3 Talent System—and it's a crucial one—is *training.* It's all about the professional development of those A Players. I'll reveal the secrets of achieving it in Chapter 3.

CHAPTER 3

Developing employees to be the best they can be

What you will learn in this chapter:

- Introduction to employee development
- What is employee engagement? And why you should care?
- Fully productive employees in 30 days: The XP3 Onboarding Training Program
- The XP3 Knowledge and Skills Assessment
- Individual Professional Development Plans (IDPs)

Introduction to employee development

Chief Financial Officer: "What if we invest in developing our employees—and then they leave us?

The Chief Talent Officer replies: "What if we don't invest in developing them—and they *stay?*"

That's what this chapter is all about. Every company grows, or dies, depending upon the ability of its key employees and managers to scale the organization. These people are the bedrock of your company; every success—and failure—begins with their abilities (or lack thereof) to do what's necessary to support the company's growth.

So take a tip from the Chief Talent Officer above: Investing in your human capital is one of the most important steps you can take to dramatically improve your organizational effectiveness, value, and growth rates. It's essential for creating a more sustainable competitive advantage.

Any company that aspires to be a market leader must invest in developing its managers—and the requisite infrastructure to support their growth. Market conditions always change, but a high-performing team creates a consistent and sustainable competitive advantage.

Yet executing on this strategy becomes exponentially more difficult as your company grows. That's why it's critical that you start as early as possible. Every business with at least three managers should implement an effective management-development program that enables its team to cultivate the professional skills they need to sustain—and grow—the company's ongoing success.

The Peter Principle, or: One role does not necessarily lead to another

The Peter Principle is a concept in management theory formulated by Laurence J. Peter and published in 1969.

We promote people to their level of incompetency.

The theory goes like this: It contends that when you select a person for a new role, you're actually choosing them based on their performance in their current role.

That's a huge difference! It's also a tragic flaw in management that gets acted out, over and over again:

- We take our top salesperson and promote them to Sales Manager.
- We elevate our top software developer to Software Manager.
- And so on.

Then, when they fail to excel in their new roles, we can't understand why. "They were so good at their former positions!" we exclaim in confusion. "Shouldn't that have made them the best possible candidates for the manager role?"

The answer is a resounding "No!" Having come along with me for the ride through all these chapters by now, I think you're savvy enough to see why. Consider, for example, what it takes to be a great salesperson:

- You need knowledge and skills that include following a structured sales process.
- You have to know how to ask great probing questions and then be quiet and listen (that's an ambivert skill).
- You need to know how to present value.

- You need to know how to overcome jections.
- You need to know how to close.
- From a behavioral perspective, you need to be independent, since you are mostly working by yourself.
- You need to possess a competitive drive, even if it's to always outdo yourself.
- You need to be assertive and impatient, which creates a natural sense of urgency to close deals in a timely manner.

And while all of this knowledge, skill, and behavior are essential for becoming a successful salesperson, *they don't even show up on the list of what it takes to be a highly successful manager.*

Don't believe me? Compare the "salesperson" list above to this "manager" list below:

- You need to know how to hire top talent who can execute on your business strategy.
- You must know how to develop employees to be their best so they produce more, stay with the company long-term, and attract other top talent.
- You need to create clear expectations, focusing individuals on those outcomes and delivering projects on time.
- You must be able to coach and motivate

employees to consistently achieve their objectives, while reducing turnover.

- You need to be able to manage by the numbers, using KPIs (key performance indicators) for constant improvement.
- You must possess high-performance team dynamics and communications skills.
- You need razor-sharp people skills and Emotional Intelligence or EQ.
- Last but not least, you need outstanding leadership skills: That's the gateway to the executive level, which entails running an area of the business with little oversight and dependence on your team to do what's necessary.

Pretty darn different, huh? Note that none of the above "manager" knowledge and skills are required to be a great salesperson, staff accountant, or software developer.

In fact, the behaviors I previously listed that define a great salesperson can actually work *against* you in a managerial role. For example, the independence and competitive streak that mark a great salesperson may make it difficult for them to function as part of a team; they may simply wind up competing against their own salespeople. And their impatience (an asset in the sales role), if not controlled, can equate to a too-short fuse, making this person quick to fire their own salespeople without first working to train and coach them.

The bottom line is this: The Peter Principle isn't just some academic theory. As this example has shown, the knowledge, skills, and behaviors required for being a great manager are completely different than what were needed for this exact same person to succeed in their previous role.

Behavioral competencies of a highly effective manager

So, how would all of these "non-Peter-Principle" management qualities show up on the MPO? You might see, depending on the specifics of the role and company:

- A balance of strategic and detailed thinking, or A = S.
- Extroversion: Someone who truly enjoys working with and through others. They'd score as trusting, optimistic, and communicative. On the MPO, that's E > S.
- They'd exhibit a balance of patience and urgency. They'd be results-oriented. And they would like/enjoy a variety of activities. On the MPO, that's A = P.
- They would show attention to detail. They'd

be compliant, conscientious, organized, and structured. That would score as S > A.

The MPO Talent Assessment

Over the years, during extensive research and development on the MPO Program, the experts at Ngenio (the company that makes the MPO) realized that individual profiles reveal a great deal beyond the "snapshot" which the assessment provides. As you know by now, the MPO reveals key personality traits and the underlying motivational needs that drive an individual's behavior.

But it gets better. The big realization was that the MPO, albeit in a more subtle and implicit fashion, can also reveal the *potential* to *develop* valuable behavioral competencies.

To be clear: It's not a crystal ball. The MPO doesn't, and can't, predict the future. Rather, it can help to define, delineate, and measure an individual's *capacity* to develop various behavioral characteristics. So it's not like the MPO will forecast the actual outcome or level of competency; instead, it can show you an individual's *inclination* to develop competencies with a certain level of ease or difficulty.

Specifically, the MPO Talent Report highlights a person's potential, as gauged against 46 of the most crucial talents for the workplace.

Talents Measured by MPO

1. Analytical Thinking
2. Applied Technical Thinking
3. Attention to Detail
4. Attentiveness
5. Capacity for Synthesis
6. Cautiousness
7. Change Management
8. Compliance
9. Conceptural Thinking
10. Consistency
11. Creativity
12. Customer Service
13. Decision Making
14. Delegation Skills
15. Empathy
16. Flexibility

17. Goal Setting
18. Helpfulness
19. Independent Thinking
21. Interpersonal Sensitivity
22. Interpersonal Skills
23. Intuition
24. Leadership Skills
25. Levelheadness
26. Listening Skills
27. Multitasking
28. Perseverance
29. Persuasiveness
30. Planning and Organizing
31. Problem Solving
32. Project Management
33. Resilience

33. Results Driven
34. Risk Taking
35. Self-confidence
36. Sense of Duty and Responsiblity
37. Sense of Initiative
38. Sense of Urgency
39. Sociability
40. Strategic Vision
41. Stress Management
42. Stringency
43. Structured Thinking
44. Supervising
45. Teamwork
46. Tolerance

Competency vs. talent

Competency is an acquired ability to produce a desired result to an acceptable standard in a specific context. *Talent*, on the other hand, refers to a natural aptitude, a predisposition, or ability to develop a specific competency.

Here's a little deeper dive on those definitions, courtesy of Ngenio:

COMPETENCY

A competency is comprised of a mixture of knowledge, know-how, and behaviors. Generally speaking, there are two types of competency:

- *Cognitive* skills: The ability to acquire specific knowledge on specific subjects.
- *Behavioral* competencies: The ability to act in an appropriate way to achieve desired results.

The MPO focuses on behavioral competencies, and measures talents in relation to the development of select generic competencies.

TALENT

Since a talent is a natural aptitude, a predisposition, or ability to develop specific competencies with greater ease, talents indicate potential.

Those are rather academic definitions. So let's bring it down to earth. Here are some examples to help you easily understand the differences between "competency" and "talent":

- A *competent* piano player is skilled and ready for the recital.
- A *talented* piano player is naturally inclined to play the piano, but may not have developed the skills yet.

Can you be a competent piano player without a talent for playing the piano? Yes, but you will work a lot harder and the road to competency will be much longer.

Can you have a talent for playing the piano but not be a competent player? Yes—and the path to competency will be easier and shorter.

Here are two more examples to help you:

- Larry Bird was a *competent* basketball player. He wasn't physically gifted, but he worked harder than anyone to develop competency.
- Michael Jordan was a talented basketball player. He didn't have to work as hard at it as Larry Bird did to achieve mastery.

If the goal is to achieve competency, then talent

helps us understand how much effort is needed to get there. I like to use the "escalator analogy" here:

As you see in the above graphic, we all start at the bottom with our raw talents. The goal is to get to the top—competency—and we have three possible paths to get there, depending on our natural level of talent.

The three levels of talent

MPO Talent Report GREEN, 7-8-9

Achieving competency comes at great ease, does not require much effort.

Talents
are reported in three categories.

MPO Talent Report YELLOW, 4-5-6

Normal range, achieving competency comes at moderate effort.

MPO Talent Report BLUE, 1-2-3

Requires effort to achieve compentency.
Like running up the down escalaor.
If you stop, you go back to where you
started and have to do it again.

Sneak peek: Actual MPO instructions and Talent Interview questions

Special thanks to the folks at Ngenio for their permission to reprint a portion here:

After selecting your key competencies above, a list of three interview questions per competency will be shown in the Interview Questions Section of the MPO Talent report for you to grade from 1 - 9 for each competency.

These questions are designed to help you establish a level of behavioral skill that you expect the candidate to have developed over time. They may also inspire further follow-up questions of your own that are more closely aligned with your specific field or industry.

It is not a viable option to assume that a natural predisposition has necessarily led to a person having actually developed the associated competency (or vice-versa). For this reason, it is wise to explicitly cross-check all required job competencies during the interview process. That said, if you notice an undeveloped or underdeveloped skill in relation to the job requirements, the natural talent level should be a good indicator of how much effort would be required to acquire the skill and perform at the appropriate level down the line. [That's the "escalator" analogy! Get it? —MM]

Sample MPO
Talent Interview questions

PLANNING AND ORGANIZING

Please tell us about the last time you had to coordinate a project or task from start to finish (plan the schedule, stages/steps, and actions) to complete it.

Please give us an example of a situation in which you predicted or foresaw a potential problem, and how you used good project planning to mitigate any fallout.

When working on a project, what methods do you use to ensure rigorous monitoring of progress, schedules, and allocated resources?

CONSISTENCY

One often faces situations that delay or jeopardize a project. Have you been through a situation in which you did not lose sight of the key objectives and showed determination and patience when faced with delays or setbacks?

Do you prefer dealing with several tasks promptly or in quick succession, or undertaking work that is medium- or long-term in length? Why?

Please discuss the last time you were unable to meet your goals due to interferences or a lack of perseverance.

ANALYTICAL THINKING

What method do you use to ensure you have all the information needed to assess a situation from every angle? Please elaborate by citing an example of your method in action.

When analyzing a situation, how do you verify or validate your hypotheses? Please discuss the last time you did this.

Please discuss a situation in which your analysis led to a faulty or flawed conclusion. In hindsight, what would you have done differently?

What is employee engagement? And why you should care?

"Employee engagement" is the hot new buzzword. Everybody's writing and talking about it.

But it's no fad. It's here to stay, for many great reasons.

The level of your employees' engagement affects just about every aspect of your organization. This includes, but is hardly limited to:

- Gross revenue/sales
- Productivity
- Talent acquisition
- Employee turnover
- Client experience
- And tons more.

So what, exactly, is it? Here's a concise definition for you:

> **Employee engagement** is the strength of the mental and emotional connection which employees feel toward their place of work.

That's all well and good, but you can't peer into their minds. You can't see those mental and emotional connections. How, then, can you recognize them? What does employee engagement *look* like? How do engaged employees act? (And don't say "Like they're about to get married!" Leave the bad jokes to me.)

Here's what you can look for, when you're looking for engaged employees:

- They speak positively about the organization, both internally and externally.
- They go above and beyond job requirements.
- They take personal accountability for the results they produce.
- They understand how their work contributes to the larger purpose.
- They're enthusiastic about what they do; they look forward to coming to work each day.
- They know the full scope of their roles, are 100-percent committed to them.
- They get excited about the challenges of their work and look for new ways of achieving the outcomes.
- They occupy roles that utilize their talents.

Employee *engagement* vs. employee *satisfaction*

These two terms get tossed around frequently together. But they're not interchangeable. Gallup research has shown that *both* relate to meaningful outcomes. But the differences between them are interesting:

- **Satisfaction** is a broad, attitudinal outcome.

It's hard to act on, and some facets of satisfaction—such as compensation, benefits, work hours, etc.—are irrelevant to performance.

- **Engagement,** on the other hand, *predicts* satisfaction, as well as many other concrete business outcomes.

Why you need engaged employees

Research from hundreds of studies has confirmed a direct link between employee engagement and business performance. But why should you care?

- In companies where 60 to 70 percent of employees were engaged, average total shareholders' return stood at 24.2 percent.
- If employee engagement fell to 49 to 60 percent, average shareholder return fell to 9.1 percent.
- If employee engagement dropped below 25 percent, the organization suffered negative total shareholder return.[5]
- Organizations with highly engaged employees achieve seven times greater 5-year total shareholder return (TSR)

---

[5] Source: Hewitt Research Brief

than organizations whose employees are less engaged.[6]

- Organizations with highly engaged employees achieve twice the annual net income of organizations whose employees lag on engagement.[7]
- Employees with lower engagement are four times more likely to leave their jobs than those who are highly engaged.[8]
- A five-percent increase in total employee engagement correlates to a 0.7-percent increase in operating margin.[9]

The manager's role in employee engagement

There is a rock-solid connection between an employee's level of engagement and the effectiveness of that employee's manager. Don't believe me? A Gallup study of 50,000 businesses found that the manager is primarily responsible for employee engagement levels. Or consider this study of global leaders, conducted by Zenger Folkman, based on a global sample of 29,906 leaders:

---

[6] Source: Kenexa
[7] Source: Kenexa
[8] Source: Corporate Leadership Council
[9] Source: Towers Perrin

Developing employees to be the best they can be

Manager's Role in Employee Engagement

In a study of global leaders,
Zenger Folkman found a direct correlation between
Manager Effectiveness and Employee Engagement

Employee Engagement by Manager Leadership Effectiveness

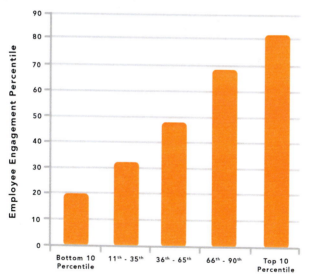

As you can see in the chart, poor leaders spawn disengagement (e.g., 20th percentile), whereas good leaders have employees who score at the 50th percentile. The employees of *great* leaders, by contrast, score above the *80th* percentile for engagement. This all adds up to a great incentive for superlative management!

And as I'd mentioned above—and as research has confirmed—engaged employees are more productive, efficient, and effective. More than 80 percent of employees come to work and perform at a satisfactory level based upon the requirements of the job. For most people, that's only *40 to 60 percent of their capacity*. Since no additional effort is required to perform the job satisfactorily, the *employee gets to decide* whether (or not!) they'll contribute that work; thus this added energy is known as "discretionary effort."

Think about that. You could nearly *double your organization's output, without adding a single person*—if you could effectively tap into the available discretionary effort. How would capitalizing on even a portion of the untapped potential of discretionary effort in your organization impact your business' bottom line?

As if you needed more convincing by this point, here's another reason you'll want to staff your organization with engaged employees: According to Gallup, highly engaged teams are *21-percent more productive* than those with low engagement. In addition, engaged employees are innovative; they always have an idea or two about what they can do better.

Engagement levers

Many factors affect employee engagement. However, the "levers" listed below have a direct cause-and-effect link to how engaged your employees are. Here's a quick overview of the essential engagement levers:

- **Job fit.** Make sure every employee possesses the three foundational strengths (Knowledge, Results, and Behavior), which we introduced in Chapter 1.
- **Relationship with manager.** People quit managers, not jobs. So make sure you have a good relationship with your employees. This isn't about making yourself lovable or popular; it's about helping your people perform. It's about properly playing your role as a supportive coach and manager.
- **Connection to mission.** For most employees, at most organizations, their first

day of work, unfortunately, begins with a visit to HR and a mountain of paperwork to complete. Do you think that that makes for an effective "engagement lever"? Of course not. Winning companies treat their new hires to an invigorating introduction by a top executive or the CEO. They talk about the history of the company, why it was created, where it's going, and how the new hire can help. You want your employees to understand that they're part of something big—and great. That's especially important with today's generation of Millennials, who truly want to change the world. It means going beyond your mission statement (assuming you have one, right?).

- **Intentional culture.** This point dovetails with the one above. Every company has its own culture, but only the best ones actively and intentionally create theirs. What kind of culture are you trying to build? Are you educating and training your people to support it? Most cultures just happen, unplanned; you should take advantage of the opportunity to plan yours.

- **Investment in growth/development.** This refers to satisfying your employees'

ongoing need to learn, un-learn, and
learn skills—and aligning those ne
with your organization's needs.

Assessing management capabilities

As I'd mentioned in the intro chapter, if you
want to hire, train, and retain top talent, you first
need to ensure that your *managers* are top talent.

How do you do that? You need a process for assessing their talents and capabilities. Once again,
we return to the three strengths of a consistent,
productive, and happy employee that we'd introduced in Chapter 1, but apply them to managers:

1. **Results:** These are specific success measurements for the role - hire 3 people by
   x date, increase productivity by x% and
   so on.
2. **Knowledge:** specific to the job that creates the desired results above. Things
   like hiring, training, and coaching for
   performance.
3. **Behavior:** The alignment of the required
   behavior of the job and the personality
   traits of the manager.

The XP3 Manager Knowledge and Skills Assessment

The XP3 Manager Knowledge and Skills Assessment is a form that was created to help you do exactly what it says it does: gauge the capabilities of your managers. It can be used as a self-assessment (employees score themselves). Or you can assess your employees. Or you can do both of the above: Use it as an assessment tool *and* a self-assessment tool.

Here's how you do it. The manager either goes through the form, topic-by-topic, with the employee; or the employee completes it on their own. Ideally, the manager reviews the responses with the employee directly. For each topic, the manager then asks that employee: "How do you think you rate?" according to the Acceptable/Unacceptable/Excellent score on the chart.

For each employee response, the manager will either agree or disagree. If they disagree, they should ask the employee to explain, with specific examples, why they ranked themselves the way they did. The manager can then use this information to facilitate a discussion addressing their differing perceptions, and what specifically is needed to attain each level in the ranking. This, then, can be used to create a development plan for that employee.

So when—or how often—should you use this tool? I recommend that you conduct an assessment when you're considering promoting someone to a management role. I also think it's a good idea to use it when a manager feels they're ready for promotion—or when you feel that an existing manager isn't living up to expectations. At a minimum, I would suggest that you administer this assessment at least once a year for each manager.

The XP3 Manager Knowledge and Skills Assessment is on the next couple of pages.

XP3 Manager Knowledge and Skills Assessment

Hiring Top Talent

Knowledge and Skills	Unacceptable	Standard	Excellent
Adheres to a structured hiring process	Does not consistently follow the process	Always	Always and is a SME (subject-matter expert)
Utilizes pre-screening tools (MPO, SQ, others)	Does not consistently utilize tools	Always	Always and is a SME
Conducts objective interviews	Does not utilize	Always	Always and is a SME
Makes objective hiring decisions (OHDs)	Does not make OHDs	Always	Always and is a SME

Develops Emploees to be Their Best

Knowledge and Skills	Unacceptable	Standard	Excellent
Utilizes valid assessment methods/tools (MPO, P3, etc.)	Does not utilize	Always	Always and is a SME
Utilizes a formal new hire onboarding program	Does not	Always	Always and is a SME
Creates an amazing experience for all new hires	Does not	Always	Always and is a SME
Educates new hires to all aspects of the company	Does not	Always	Always and is a SME
Helps new hires build relationships with as many other employees as possible	Does not	Always	Always and is a SME

	Unacceptable	Standard	Excellent
Facilitates training programs	Does not	Occasionally	Always and is a SME
Creates Individual Development Plans (IDPs) for all new hires	Does not	Yes, but does not utilize	Always and is a SME

Clear Expectations

Knowledge and Skills	Unacceptable	Standard	Excellent
Establishes clear written expectations for all employees	Does not	Yes, but does not hold accountable	Always & Holds Accountable

Coaching and Motivating

Knowledge and Skills	Unacceptable	Standard	Excellent
Utilizes the MPO to identify intrinsic motivational needs	Does not	Occasionally	Always and is a SME
Conducts a regularly scheduled 121	Does not	Yes, but >monthly	Once or More per Month

Managing by the Numbers

Knowledge and Skills	Unacceptable	Standard	Excellent
Utilizes metrics wherever possible	Does not	Occasionally	Always and is a SME
Focuses employees on metrics	Does not	Occasionally	Always & Achieves Goals

Team Dynamics and Communication Skills

Knowledge and Skills	Unacceptable	Standard	Excellent
Utilizes the MPO communication tool to identify intrinsic styles	Does not	Occasionally	Always and is a SME

Facilities team dynamics training	Does not	Occasionally	Always and is a SME
Models effective communication	Does not	Occasionally	Always and is a SME

Emotional Intelligence (EQ)

Knowledge and Skills	Unacceptable	Standard	Excellent
Self-awareness	None	Some	Yes
Self-regulation	None	Some	Yes
Empathy	None	Some	Yes
Social Skills	None	Some	Yes

Leadership

Knowledge and Skills	Unacceptable	Standard	Excellent
Leads by example	Never	Occasionally	Always
Team Turnover	High	Normal	Low
Team Morale	Low	Normal	High
Team Production	Low	Average	High

The Gallup Q¹² Survey

Another great method for assessing how well a manager and the company are addressing employee engagement is by administering the Gallup Q12 Survey. Gallup researchers spent decades writing and testing hundreds of questions, because their wording and order mean everything when it comes to accurately measuring engagement. This was published in the best-selling book, *First, Break All The Rules: What the World's Greatest Managers Do Differently*.

Their research yielded 12 essential questions; when answered, they comprise one of the most effective measures of employee engagement and its impact on the outcomes that matter most to your business.

To administer the Q12, have your employees answer 12 simple questions, each of which relates directly to performance outcomes. Scores are based on a 1 to 5 scale with 1 = strongly disagree and 5 = strongly agree.

THE GALLUP Q¹² SURVEY

1. I know what is expected of me at work.
2. I have the materials and equipment I need to do my work right.

3. At work, I have the opportunity to do what I do best every day.
4. In the last seven days, I have received recognition or praise for doing good work.
5. My supervisor, or someone at work, seems to care about me as a person.
6. There is someone at work who encourages my development.
7. At work, my opinions seem to count.
8. The mission or purpose of my company makes me feel my job is important.
9. My associates or fellow employees are committed to doing quality work.
10. I have a best friend at work.
11. In the last six months, someone at work has talked to me about my progress.
12. This last year, I have had the opportunity at work to learn and grow.

Visit Gallup.com to learn more.

Fully productive employees in 30 days: The XP3 Onboarding Training Program

A few years ago, I was approached by one of the country's largest window-replacement contractors. I helped them to hire and train 18 new salespeople.

From that group, 17 graduated the onboarding program—and then went on to sell over $1.4 million in
replacement windows in their first 30 days in the
field!

By any measure, that's a stunning accomplish-
ment. How did we pull it off?

First, we hired the right people, using the XP3
Talent hiring process. But we didn't stop there. We
then created an *amazing experience* for them, using
the XP3 Onboarding Training Program—the topic
of this section of this chapter!

You, too, can achieve results like this. That's what
this program is designed to do. Our system will:

- Shrink your new hires' learning curve.
- Get them fully productive in the shortest
  time possible.
- Assimilate them into your organization's
  culture.
- Make them part of the "team."
- Build relationships with coworkers.

What does this all lead to? The retention (oops,
that's Chapter 4, but you'll forgive me for getting a little ahead of myself) of productive, happy
employees!

Like most things XP3, the Onboarding Training

Program utilizes the power of "three." Specifically, it includes three objectives you need to focus on, in order to get your new employees fully productive in the shortest amount of time:

The XP3 New Hire Onboarding Objectives

Here are the three (yes, *three*) things you'll need to accomplish:

1. **Create an amazing experience.** You want to wow your new hires!
2. **Build relationships** with as many of your other employees as possible.
3. **Train them on *all* aspects of your company**—not just the role you hired them for.

Let's look at each of these topics in detail:

How to create an amazing experience for your new hire

First—and this may sound like a no-brainer, but you'd be surprised by how many companies fail to do it—*prepare for that new hire in advance.* Be organized. Use a checklist. Address everything, and I mean *everything*, prior to their arrival:

- Clean their workstation. Use disinfectant.
- Clean their company vehicle. Fill it with gas.
- Check and clean all their equipment. Make sure everything's in working order.

Let me put it this way: Have you ever started a new role, only to discover, on Day One, that you had to clean out the last person's stuff? Were you the one who had to wipe that don't-wanna-touch-it sticky stuff off the phone? That's not a very welcoming feeling, is it?

Next, call in the Welcome Wagon. Notify everyone of the new employee's arrival; make sure they go out of their way to make them feel welcome.

Again, consider the flip-side as motivation: Have you ever started a new job, only to be greeted by a receptionist who 1) had no idea who you were, 2) where you were supposed to go, and 3) wasn't very friendly ? How "welcoming" did that feel?

Finally, *don't* start that new hire in HR, filling out paperwork! In a word: *Boring!*

Instead, begin their first day with a tour. Introduce

them to people. Have the President or CEO give them a short (15- to 20-minute) overview of the company, one that addresses:

- How the company got to where it is today.
- The exciting vision for the company.
- The new hire's role in helping the company to achieve that vision—and how their role (like all roles) plays a critical part in achieving that success.

In case the CEO is unavailable to make a live presentation, record a video.

Build relationships

This is Number 2 of my three-point checklist above. Get as many people as possible involved with the new hire's training. Make sure that *they* understand the three objectives on the checklist.

Your employees should begin by telling the new hire about themselves:

- How long they've been with the company.
- What their role is.
- What they like about working here.
- A little about their personal life—what they do when they're not working.

Then flip the tables. Have the new hire do the same.

Keep it it cordial. Schedule lunches and/or happy hours with the others for the first few days to accomplish this. It will—and should—be fun!

Train them on all aspects of your company

Remember: You want them to learn all about your company—not just the role you hired them for. This does *not* mean that you need to train them on how to do everything. Simply explain to them what each department does, who the key players are, and how their work supports what the new hire will be doing. Here's how you do it:

First, create an onboarding training schedule. It should be a day-by-day, hour-by-hour schedule, detailing what the new hire will learn, and who will teach it. It should feature an introduction to all departments and their key players, including what they do, and how that supports their department. It should also include a checklist, to ensure that the new hire has learned and retained the information.

Once you've created this schedule, email it to the new employee in advance; it will tell them what to expect, while letting them know that you're organized.

After they've completed the schedule, then you can focus on the training them on the specific job for which they were hired.

The end-of-day check-in

At the end of each day of the first week, ask the new employee what went well, as well as what the company can do better, to improve their experience. Take notes.

The obvious reason to do this is to learn more and continuously improve your onboarding. But it's also to create company advocates. In fact, your goal is to have that employee go home and *tell everyone they know about the great decision they made by joining your company.* That's key to building their engagement.

A client of mine recently used the XP3 Talent System to hire a new architect, project manager, salesperson, and lead carpenter—and they all started on the same day. At the end-of-day check-in, the lead carpenter was positively gushing: "I called my wife after lunch to tell her what a great company this is! I've been in this business for 25 years, and I've never been treated so well! I thought I'd just show up, fill out a ton of paperwork, and then be shuffled off to the job site. But this was just amazing! You all made me feel welcome! I've already made new friends, learned a lot, and am totally excited about my future here!"

That's the kind of reaction you want to create. Your goal is to have engaged employees who are fully productive and integrated into your company's culture within 60 to 90 days.

The XP3 Onboarding Process Checklist (OPC)

Onboarding a new hire includes a lot of steps. You don't have to memorize any of these; simply download the free XP3 Onboarding Process Checklist—a ready-to-use Excel sheet that you can easily customize for your needs—from XP3Talent.com.

The Checklist includes a lot of pre-populated tabs, such as those for HR, Supervisor, Onboarding Schedule, etc.

On-Boarding Process Category Tabs

New Hire Process Overview

New Hire Info

Talent Acquistion

Human Resources

IT-Help Desk

Supervisor

Marketing

On-Boarding Schedule

Each tab contains a complete checklist for each department to help ensure that you address all critical items. Here, for example, is a snapshot of some of the items in the IT/Help Desk tab:

XP3 Example of the IT Help Desk Checklist

Pre-Hire Tasks	Description
Start Date	Date employee will need resources available
New Hires Name	First, Middle, Last as it will appear in our directory
Job Title	Official Title on business card if applicable
Office Location	Where we need to set up workstation & phone
Workstation Requirements	Desktop, laptop, printer
Cell Phone	
Application Needs	Pre-installed software needed
Phone set up	Standard or CS agent
Voicemail set up	
Email set up	first initial, last name@company.com with signature set up
Group email set up	What groups to assign All, Managers, Field Marketing, Sales, etc.
WNA Access	WNA access required? Which restricted directories if any?
Salesforce.com set up	
iPad set up	
Web brower security limitations	Does this employee need access to special sites our firewalls may block

New Hire Info | Talent Acquisition | Human Resources | IT-Help Desk

153

Here's another example to whet your appetite: a sample of the XP3 OPC Training Schedule:

XP3 On-Boarding Process Checklist

Training Plan			**Employee:** James Doe			**Position:** Sales Manager		
Week 1								
Wednesday	**Trainer**	**Task**	**Thursday**	**Trainer**	**Task**	**Friday**	**Trainer**	**Task**
8:00 AM	BM	Welcome & Tour	8:00 AM	CM	Field Time	8:00 AM	BM	Check in
9:00 AM	CM, KL	Ops Meeting				9:00 AM	NK	Folder Process
10:00 AM	BE	Sales Meeting				10:00 AM	NK	Scheduling
11:00 AM	BE	Sales Meeting				11:00 AM	NK	Financial Board
12:00 PM	Leaders	Lunch	12:00 PM		Lunch	12:00 PM	KZ, NK	Lunch
1:00 PM	BN	Admin/Intake	1:00 PM	BC	Field Time	1:00 PM	KZ	Field Time
2:00 PM	BM	P3 & Overview						
3:00 PM	PM's	Field Time Production						
4:00 PM	PM's	Field Time Production						
5:00 PM	BM	Day 1 Wrap Up	4:30 PM	KL	Day 2 Wrap Up	4:30 PM	BM	Day 3 Wrap Up

Week 2								
Monday	**Trainer**	**Task**	**Tuesday**	**Trainer**	**Task**	**Wednesday**	**Trainer**	**Task**
8:15 AM	BM	Planning	8:00 AM	BE	Sales Manager Role	8:30 AM	CM, NK	Ops Meeting
9:00 AM	Leaders	Big Rocks				9:00 AM		
10:00 AM	MR, MS	Accounting				10:00 AM	BE	Sales Meeting
11:00 AM	MR	Accounting						
12:00 PM	MR, BN, TR	Lunch	12:00 PM	BE, BP	Lunch			
1:30 PM	KL	Managers Meeting	1:30 PM	BP	Sales Ride-Along	1:00 PM	RC	Lunch
2:00 PM	KL	Managers Meeting						
3:00 PM	TR	Design/Drawing						
4:00 PM	TR	Design/Drawing						
4:30 PM	BM	Day 4 Wrap Up	4:30 PM	JG	Day 5 Wrap Up	7:30 PM	BM	Day 6 Wrap Up

On-boarding Schedule | Wednesday | Thursday | Friday | Monday | Tuesday | Wednesday(...)

Ongoing skills assessment and professional development plans

"Would you tell me, please, which way I ought to go from here?"

"That depends a good deal on where you want to get to," said the Cat.

"I don't much care where--" said Alice.

"Then it doesn't matter which way you go," said the Cat.

—From *Alice's Adventures in Wonderland*,
by Lewis Carroll

You wouldn't start a project, Alice-like, without any idea of where you want to go. Nor would you take any random route to achieve it, as the Cheshire Cat advised. Why, then, would you approach the skills-development and assessment of your new hires any differently? A solid plan, with explicit goals, will provide structure and focus, ensuring that you make the most of this opportunity.

The following is an example of the detail page for a typical training session:

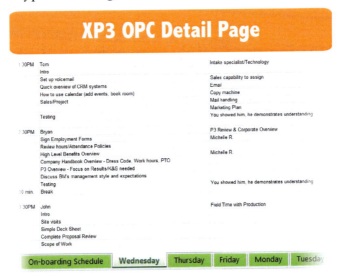

The XP3 OPC Detail Page

1:30PM	Tom	Intake specialist/Technology
	Intro	
	Set up voicemail	Sales capability to assign
	Quick overview of CRM systems	Email
	How to use calendar (add events, book room)	Copy machine
	Sales/Project	Mail handling
		Marketing Plan
	Testing	You showed him, he demonstrates understanding
2:30PM	Bryan	P3 Review & Corporate Overview
	Sign Employment Forms	Michelle R.
	Review hours/Attendance Policies	
	High Level Benefits Overview	Michelle R.
	Company Handbook Overview - Dress Code, Work hours, PTO	
	P3 Overview - Focus on Results/K&S needed	
	Discuss BM's management style and expectations	
	Testing	You showed him, he demonstrates understanding
10 min.	Break	
3:30PM	John	Field Time with Production
	Intro	
	Site visits	
	Simple Deck Sheet	
	Complete Proposal Review	
	Scope of Work	

| On-boarding Schedule | Wednesday | Thursday | Friday | Monday | Tuesday |

The XP3 Knowledge and Skills Assessments

Remember the XP3 Manager Knowledge and Skills Assessment, which we had discussed earlier in this chapter? Well, you can easily create "spin-off versions" of it to suit the requirements of most other roles. For example, when I first needed to hire talent-acquisition specialists for my company, I created a Talent Acquisition Specialist Knowledge and Skills Assessment. For one of our construction clients, we helped create a Production

Manager Knowledge and Skills Assessment. Here's a quick excerpt of that one:

XP3 Knowledge and Skills Assessment: Project Manager

Performance Level Breakdown
What is the expectation?

Safety	"Excellent"	"Standard"	"Unacceptable"
PPE (including Fall Protection)	Always leads by example.	Wears & encourages the use.	Does not wear or encourage.
Driving	Impeccable driving record.	Adheres to all traffic laws.	Minor infractions.
Cellphone use while driving	Will not use, period.	Will not use / uses autoreply.	Will openly use.
Jobsite Housekeeping	Impeccable jobsites.	Reminds and encourages.	Won't address issues.
Tool & Equipment	Ensures self and others are using safe equipment.	Uses tools and guards properly.	Will cut corners on safety.
Sets good example	Leads by example.	Will practice what he preaches.	Does not set example to follow.
Holds effective jobsite meetings	Leads and drives safety meetings.	Conducts a safety meeting.	Rarely holds meetings and goes through the motions.

Quality	"Excellent"	"Standard"	"Unacceptable"
Quality of own work	Takes great pride in work.	Gets work done with minimal mistakes.	Does the bare essentials.
Inspecting material	Confirms every piece is there and good.	Inspects the load.	Does not check the load.
Inspecting workmanship	Holds crew accountable for the highest level of quality.	Walks the job for correctness & requires the crew to make repairs.	Does not inspect and allows poor quality to slice by.
Performs Zero Defects Checklist	Uses forms for training tool as well.	Completely & Thoroughly.	Rarely turns them in.
Fixing deficiencies	Catches mistakes before inspectors and customers.	Make repairs / orders repairs.	Ignores poor quality.

Project Management Advancement   **Skills Assessment**   ⊕

The bottom line is that you can create one of these for virtually any role to help identify an individual's strengths and areas for improvement. The great thing about this tool is that it actually gets your employees to self-manage their own development and career progression.

Here's how you do it. Once you've created the new assessment, the employee can utilize it to monitor their own progress and advancement. For example, when they believe they're ready for a review (and possible advancement), they come to you, their manager, and utilize the particular assessment you've created for their role to review and discuss their perceived progress.

Thus you both have a tool for facilitating the conversation and defined parameters to use as a gauge for how effective they are vs. the typical subject approach. You also have an opportunity, as a manager, to either agree and confirm, or disagree and discuss what specifically is missing and/or needed to achieve the desired level of performance.

The career advancement chart

Over the years, we've also helped our clients to create advancement charts; these outline specifically what is required for a person to advance to the next level in their role.

Take a look at the screen shot from the example that follows. It shows the potential career advancement for a Project Manager, as they move up the ranks in skills and seniority. The abbreviated column headers (such as "APM1," etc.) stand for continually more demanding positions of responsibility, such as "Assistant Project Manager 1," "Assistant Project Manager 2," "Assistant Project Manager 3," then "Project Manager 1," "Project Manager 2", and finally "Senior Production Manager" (who manages Project Managers).

Developing employees to be the best they can be

XP3 Career Progression Chart

Project Management Skills & Responsibilities	APM 1	APM 2	APM3	PM 1	PM 2	SPM
	o	o	o	o	o	o
Planning						
Provide weekly plan via Cloud or Spreadsheet (daily updates required)	x	x	x	x	x	x
Folder review		x	x	x	x	x
Hand-off with sales: facilitate the definition of project scope, goals and deliverables			x	x	x	x
Drawing Review	x	x	x	x	x	x
Take-off: checks estimators accuracy		x	x	x	x	x
Define project tasks and resource requirements			x	x	x	x
Order Acknowledgement		x	x	x	x	x
Assign crew				x	x	x
PO's for all work on job				x	x	x
Plan and schedule project timelines (Dumpsters, materials, crews)			x	x	x	x
Order materials		x	x	x	x	x
Check & Evaluate Planning Skills of Others					x	x
Monthly Forecasting (Calendar & Financials)						x
Pre-Construction Process (with consumer)						
Verify client contact info (phones, emails, etc.)		x	x	x	x	x
Review scope of work & clarify - obtain signature		x	x	x	x	x
Review drawing & clarify - obtain signature		x	x	x	x	x
Review site and confirm tech measurements		x	x	x	x	x
Ask customer to define job as they were sold it			x	x	x	x
Discuss timeline & calendar (define steps for daily & entire job)			x	x	x	x
Review payment schedule - modify as necessary			x	x	x	x
Establish best method of communication			x	x	x	x
Establish expectation & boundaries for communication			x	x	x	x
Establish face to face meeting shedule			x	x	x	x
Establish payment receipt expectations			x	x	x	x
Discuss AWO & CO (provide examples & define process)			x	x	x	x
Discuss Warranty (& Warranty Process) - obtain signature			x	x	x	x
Open discussion regarding job ("have we covered everything as sold?")				x	x	x

Project Management Advancement   Skills Assessment   (+)

The completed advancement chart, conveniently, becomes the basis of the Individual Development Plan or IDP.

The Individual Development Plan (IDP)

What, exactly, is an IDP? It's a written plan for developing the knowledge, skills, and competencies required to support the employee's needs and goals—and the organization's objectives.

The IDP outlines the employee's career goals, and the steps they must take to achieve them. It helps them focus their professional development by providing an "action plan" for skill development and career management. It also gives employees a way to document their development via assessment and reflection, empowering them to accelerate their growth. As such, it's a great tool to help them—and you—to identify, organize, and plan for the next stage.

The five key tasks for creating an IDP

Since the IDP is a totally custom tool, I can't offer you one for download from the XP3 Talent website. But I can help you to make one for yourself.

Don't be intimidated. It's just a Word doc. To create yours, simply address the following five topics:

1. **Align the employee's development needs with your company's business needs.** What results must the employee achieve in order to support the company's objectives? What activities or tasks are required to achieve those objectives? Once they've answered that question, have them identify the necessary

knowledge, skills, and competencies that support those goals.

2. **Assess knowledge and skills.** Have the employee list their skills and knowledge. Ask them to identify the strengths they've acquired. Which areas are sufficiently developed? Which others would benefit from additional learning? They should make note of both.

3. **Set goals.** Have them write down their professional and personal goals. Have them ask themselves: "Where do I want to go? What areas of my work do I wish to develop?" Have them write down the overall goals they want to accomplish in the short-term (this year), mid-term (next one to two years), and long-term (three to five years). Ask them to determine how their goals align with their knowledge and skills—or if any mid-course corrections are required.

4. **Create an action plan.** Ask the employee to determine how they're going to get to where they want to go. (Remember Alice and the Cheshire Cat!) Have them write down the skills and knowledge they want to develop. Then have them identify strategies or action steps they'll need to take to achieve their goals. It's helpful to have them create a timeline for starting and completing work on their stated goals.

5. **Document their development.** Finally, have them track their development by charting their accomplishments. Let them use personal statements, as well as feedback from their supervisor, peers, and other employees. Make sure that they revisit their goals occasionally to determine whether (or not) they need to develop additional skills or knowledge to achieve them.

You aced the trainer training!

You just learned a ton. You understand the importance of employee development. You can wittily mention the Peter Principle at cocktail parties. You understand the subtle differences between "competency" and "talent," as well as those between employee "engagement" and "satisfaction." You've learned about those all-important employee-engagement levers, and how to use them. You've learned how to create an amazing onboarding experience for your new hires. You've learned about assessments, career progression, and IDPs.

That's a lot. You should be justifiably proud. And you should also be excited, because we're just about to embark on the last phase of our journey: Retention. How do you hold on to those A Players you've worked so hard to hire and train? That's the subject of Chapter 4.

Performance coaching: The key to retention

What you will learn in this chapter:

- Introduction to performance coaching
- Micromanagement vs. proactive management
- Define and manage clear outcome-based expectations: Introducing the XP3 Peak Performance Profile™ (P3™)
- How to identify performance gaps and corrective actions
- The key to enhanced communication and motivation
- Guidelines for conducting effective one-to-one's

Performance coaching

In a book called *"Hire, Train, Retain,"* you would certainly expect the concluding chapter to be about retention. So why am I starting off by talking about "performance coaching"? Well, it turns out that you can't have one without the other.

People don't quit jobs; they quit managers.

These days, providing a paycheck no longer suffices to make you a "good" employer. As we'd discussed in Chapter 3, employee engagement is a key element to retaining top talent. You must make a conscientious effort to create a positive employee experience, in order to retain that top talent you worked so hard to find, hire, onboard, and train.

How do you do this? The primary "ingredient" in this "recipe" is *emotionally intelligent leaders*—the kind of people who can understand and facilitate that outcome. Contrast that to bean-counting supervisors who only look at the numbers, and are only good at dictating directives. Think about it: Nobody wants to be managed. But almost everyone is open to being coached. Therefore, in today's world, if you want to be a highly effective manager, you need to become a veritable performance coach.

Coaching is a more collaborative process than merely managing. It builds closer relationships. It celebrates success. It looks forward, optimistically, to seeing what's possible, and how to get there. And it's highly goal-oriented—the antithesis of that strange loyalty which people unfortunately develop to rote, trivial tasks.

Climbing the "leadership staircase"

We begin our careers as individual contributors. Next on the corporate ladder is Manager. The ultimate destination is *Leader*.

What, then, is the difference between those last two? What separates a leader from a manager?

For starters, a manager manages individual contributors; a leader manages managers.

I know what you're thinking: "Does being a manager preclude being a leader?" No. Not at all. Consider my definition of "leadership":

> *Leadership: The ability to consistently get results from others in a positive manner.*

Most people confuse leadership with management; they believe they're a leader because they were given a "manager" title. But it takes more than a title to become a true leader.

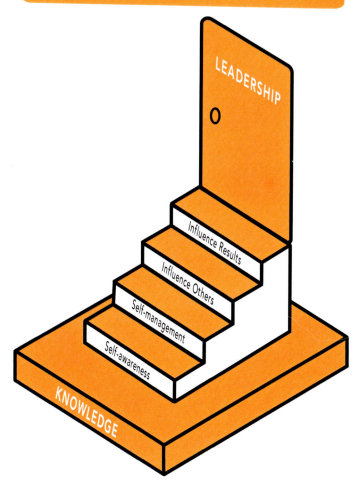

Leadership Staircase
The Path to Achieving Business Results

LEADERSHIP

Influence Results

Influence Others

Self-management

Self-awareness

KNOWLEDGE

This is going to sound pretty Zen-like, but follow me on this: *The path to achieving business results and the path to leadership are the same path.* This bears repeating. So I'll repeat it:

The path to achieving business results and the path to leadership are the same path.

This path is built on the foundation of knowledge. Take a look at that "staircase" illustration on the previous page. You can gain knowledge in various ways:

- You could read a book, just like this one.
- You can attend a formal training program, such as the XP3 Leadership Academy.
- You could even watch some YouTube videos.

There are many ways to learn; the key is to always be learning. That said—and with knowledge as your foundation—if you want to climb the leadership staircase and ascend to the level of being a true leader, you'll need to climb the following steps:

The first step is **self-awareness**. You need to be self-aware that you must make a change in your approach or behavior. For example, a particular task might require you to be more detailed. So first, you must possess the *knowledge* that this task requires you to become more detailed. Then, you need to know how to become more detailed, in order to accomplish it. That's a mini journey in self-awareness.

The second step is **self-management**. Have you ever met someone who was incredibly *un*-self-aware? (Think: "Bull in a China shop.") How good was that person at self-management? You can't self-manage without being self-aware. The first is a prerequisite for the second. In other words: You can't skip any of the steps!

The third step is the ability to **influence others**. Once you've mastered self-awareness and then self-management, you can then learn to effectively influence others' behavior and actions.

The fourth step is **influencing results**. At this level, you can now begin to purposely impact the results of those you lead. This is what leadership is all about. Remember: *Leadership is the ability to consistently get results from others in a positive manner*.

Managing people requires enthusiasm

"Being a manager would be great, if it weren't for all those people knocking on my door and bothering me all day!"

I'll bet you've heard people say this line, or something similar. You may well have thought it to yourself on several occasions. That's okay; being a manager (or leader) is not for everyone.

Most people get into management because they believe it's the only path to achieving what they want from their work life. And what do they want? Typically it's things like opportunities to learn, recognition for their efforts, additional challenges, and, of course, the opportunity to make more money.

The word "believe" is really important in what I just described. Unfortunately, a long time ago (probably before you and I entered the work world), a hierarchy was put in place that provided only one path that allowed people to get those things. And that path led straight to—you guessed it—management. Thus the "because they believe" qualifier I used above.

Here's good news: The world has changed. Today, there are multiple paths for people to get all those things that they want, *without* having to become a manager. Of course, *that* is based on the requirement of having emotionally-intelligent leaders who can think outside the box, and creatively provide those "want-fulfilling things" for their employees.

I think you can see where I'm going here: To be a

highly effective leader, you must *want* to be a highly effective leader—and for the right reasons. These include:

- Truly enjoying working with others.
- Honestly caring about others' well-being.
- Deriving great satisfaction from helping others achieve their objectives.
- Seeing those people who knock on your door as an opportunity to coach and mentor—and not a nuisance or interruption.
- Getting excited about the potential you can achieve in helping others.
- Enjoying the organization of work processes.
- Accepting responsibilities for yourself and your employees.
- Being a great listener.
- Possessing excellent communication skills.
- Having good decision-making skills.
- An easy ability to accept differing opinions.
- An equally strong facility for effectively resolving conflicts as they arise.
- The ability to keep calm, even in stressful situations.

If these are qualities that you can relate to, and/or aspire to, then you may just have what it takes to be a highly effective leader.

And what if this list doesn't resonate with you? Take heart; it's not the end of the world. Don't try and force yourself into the role. That said, you will need to identify someone else in your organization who will be a good fit and thus able to take the lead. You can then find a role that takes advantage of your unique talents.

Micromanagement vs. proactive management

More times than I can count, I've asked management candidates to describe their style to me, and the first thing they'll say is, "Well, I'm *not* a micromanager."

Red flag alert! I've found that when they say this, the real answer is either 1) they *are* a micromanager (read: "Denial"), or 2) they're what I call an "absentee manager," *i.e.*, one who's completely hands-off. Which is just as bad.

Being a performance coach means being a proactive manager. There's an old saying I love:

> *What gets tracked and measured gets done. What doesn't get tracked and measured, can't be managed.*

Kraig Kramers, author of CEO Tools and a highly

effective turnaround CEO, has said that this was the key to his success. Every business has key performance indicators or KPIs, and a highly effective manager/leader knows the specific KPIs for their business/department, and for every person they lead. That's because when you understand these KPIs—these *measurements*—you can proactively manage the outcomes. Your understanding should also encompass the frequency of the intervals of each KPI, and therefore how frequently you need to monitor, and take action on, them. Kramers has also said that you should never wait more than 30 days on any KPI: that would put you in a position of continually trying to play catch-up. And you certainly don't want to do that.

Define and manage clear outcome-based expectations: Introducing the Peak Performance Profile (P3)

One of the primary reasons that employees fail to succeed in their jobs is that employers fail to clearly define "success" for that role. In most workplaces, in fact, there's a stark disconnect between how the *employee* defines success, and how that employee's *supervisor* defines the exact same thing.

Don't believe me? Try this experiment:

Ask your employees to make a list of their work priorities, along with the due dates for each of these tasks. Meanwhile, you (the supervisor) should independently create the same type of list for your employee's job. After you both complete your lists, get together and compare them. What do you see? Don't be shocked: For most employee/supervisor relationships, you'll probably find some pretty big gaps. This is one of the main reasons that annual reviews often degenerate into unresolvable he-said/she-said dialogs—it's just a confrontation of subjective opinions!

Put it this way: If you're running a race and don't know where the finish line is, how will you know when you're done?

Not to worry. Help is on its way. The XP3 Peak Performance Profile, or "P3" for short, will provide you with an objective and scientific approach to defining exactly what's required for a job. Even better: It's a free download, available at XP3Talent.com.

The Peak Performance Profile was built based on the the three strengths that create a consistent, productive, and engaged employee. By this point in the book, you probably have them memorized already, but here they are regardless:

1. Application of that knowledge into a set of skills that creates **results**.
2. Specific **knowledge** of the job.
3. Alignment of the required **behavior** of the job with the intrinsic *personality traits* of the individual.

The P3 breaks down job tasks by category. It lists the knowledge and skills required for those tasks, and the ideal and/or required behavioral traits (attention to detail, assertiveness, communication style, etc.) for those tasks, as well as the estimated percentage of time needed for doing those tasks—and most importantly, the results ("How success will be defined") in doing the tasks.

THE XP3 PEAK PERFORMANCE PROFILE

An objective and scientific approach to defining exactly what is required in a job. The P3 breaks the job into the following five categories:

1. Tasks by category
2. Results/success measures
3. Knowledge and skills
4. Required behaviors
5. Estimated percentage of time to be spent in each task category

The P3 is not a single-purpose tool like the MPO; it's more of a Swiss army knife. You can use it:

- **As a hiring tool** to help you focus your interview on the required knowledge and skills—and to help the candidate understand the full scope of the role. This, in turn, will help them decide if they can, and want to, do the job.
- **As an assessment tool,** to gauge an employee's knowledge and skills as they pertain to the role's requirements.
- **As an aid for creating an individual development plan (IDP)** for that employee, based on the above assessment.
- **As a performance-management tool,** empowering you to coach the employee on the appropriate results and success measurements.

How to create the P3

Here's some news you'll like: Regardless of the role you're targeting, the process for completing a P3 is identical. Whether you're trying to fill a position for Director of Sales, Project Coordinator, or even Chief Operating Officer, simply visit Xp3Talent.com and download the free P3 template. Then follow these steps:

P3 Example of Task Category 1 for a Sales Manager

Task 1: Sales Performance Management	Required Behavior:	% OF TIME
1. **Develop/Demonstrate Expertise** – Develop /maintain an exceptional understanding of the industry, company's offerings, strategy, culture and values to function as hands on as well as strategic leader and manager of the revenue generating function.	Strategic/ Tactical	**15%**
2. **Hands on Balancing of Workload** - Develop /apply deep competence in CRM to assign and reassign (workload balance) new /old prospects and accounts *ensuring each is receives the level/quality of attention needed to drive sales/up-sell.*	Social/Things	
3. **Monitor Closely** - Proactively track, monitor and manage towards the achievement of the increased sales and profitability goals & objectives of the company.	Proactive/ Methodical	
4. **CRM Management:** Get "hands dirty" daily in the CRM finding examples of excellence (or areas of coaching need) in: 1) adherence to process, 2) professionalism, accuracy and consistency with marketing messaging, 3) evidence of strategy in all sales team members approach as well as conduct an ongoing assessment of the sales process/CRM itself.		
5. **Inspect what you Expect** - Ensure all communication verbal and written between members of the sales team and external parties is accurate and consistent with policies and the brand, upholds the marketing messaging and reflects exceptional professionalism.		
6. **Coaching/Mentoring:** Conduct weekly 1 to 1 meeting with your sales managers to discuss progress, next steps and provide the necessary support to help them achieve their goals and objectives with their individual sales teams.		

Results/Outcomes:	Required Knowledge and Skills:
1. Deliver formal presentation to management team to demonstrate your understanding of our industry, company offerings within 30 days of your start date.	1. At least 3 years' experience successfully managing, building, training, retaining and mentoring a high-performance sales team
2. Meet or exceed the aggregate sales team goals.	2. Demonstrated experience in coaching and developing sales managers and reps to achieve a greater individual and team outcome
3. Ensure weekly one to ones are conducted without exception.	3. Demonstrated experience working in a CRM system

Step One: Identify and categorize the tasks

Using the template you just downloaded—or even a blank sheet of paper—make a list of all the tasks for the role. Let's say you're trying to hire a director of sales. Ask yourself: "What tasks is that person responsible for?" Write down your answers.

Next, you'll want to *categorize* these tasks. Here, you'll be creating different headings. Look at the tasks on your list. For our "director of sales" example, you might see entries such as:

- "Coach and mentor sales team."
- "Determine individual sales quotas."

Those two seem related; what might be a good heading that would encompass them both? How about "Performance coaching"? Sure, that should be good.

Keep going. Categorize the rest of the tasks. Other possible headings for "director of sales" might be:

- "Hiring and onboarding."
- "Sales training."
- "Sales administration."

Add this to the "Performance Coaching" header which you've already created, and you're up to four "Task" categories.

Now enter these headings, and each heading's related tasks, into the P3 template.

Example of VP of Sales Headings:

P3 Task Categories

Task 1	Sales Performance Management
Task 2	Assess & Develop Your Sales Team
Task 3	Hiring & Retention
Task 4	Sales Strategy & Marketing
Task 5	Administration & Reporting
Task 6	Company Leadership

Step Two: Identify the results

For each task category, try and identify as many *results,* related to each of those tasks, as you can. Use the SMART formula to ensure that each goal is as specific as possible:

SMART Goals

S   Specific

M   Measurable

A   Achievable

R   Relevant

T   Time-Bound

So, for example, for the task "Coach and mentor sales team," the "Result" could be: "Meet or exceed aggregate sales quota of the team, which is $7M in annual sales." Another "Result" might be "Identify and launch a sales training program by the end of your first 90 days on the job."

P3 Results/Outcomes

provide the necessary support to help them achieve their goals and

Results/Outcomes:
1. Deliver formal presentation to management team to demonstrate your understanding of our industry, company offerings within 30 days of your start date.
2. Meet or exceed the aggregate sales team goals.
3. Ensure weekly one to ones are conducted without exception.

Step Three: Identify the required knowledge and skills

Proceeding category-by-category, now list the *knowledge* and *skills* required for each task.

For example, if the "Task" is "Coach and mentor sales team," and the "Result" for that task is "Meet or exceed aggregate sales quota of team," ask yourself: "What knowledge or skills would be necessary for the director of sales to achieve that result?" It's probably "Some experience with formal training programs."

Similarly, if the task is "Working with your team to develop individual sales plans," and the optimal "Result" might be: "Have written sales plans completed for each rep within 30 days of their hire," then naturally the required skill for this task would be "Demonstrated experience developing sales plans." Pretty straightforward.

Step Four: Identify the required behaviors

Take a look at the "Behavior" column. To work with it, we're going to use—yet again—the three primary behaviors that we've been talking about since Chapter 1:

P3 Primary Behaviors

1. Decision-making Style

A>S = Generalist

This is the strategic-thinking/big-picture person. They're focused on the bottom line. They believe that "Close is good enough," and that "There's always a Plan B."

S>A = Specialist

This is the tactical thinker. They like to do things right the first time. They're precise, exact, detailed, and organized.

2. Social Orientation and Communication Style

S>E = a "Thing-Oriented" Person

This personality prefers to work with with things, vs. people. They're cautious, slow to trust, introspective, and analytical. Their communication style is direct, factual, and straightforward.

E>S = a "People-Oriented" Person

They like to work with and through people. They're quick to trust. They draw energy from others. They think out loud. Their communication style is friendly, open, and persuasive.

3. Work Pace

A>P = Faster

This person is results-oriented. They work at a fast pace, and evince a sense of urgency in all they do.

P>A = Slower

This person is process-oriented. They're patient, methodical, and prefer a slower, steady pace.

Now, look at each *task*, and think of the *behavior* that would best suit it:

- **Decision-making style:** For each task, ask yourself whether it requires more *strategic* or *tactical* thinking. Then label it as "A>S," or "S>A," accordingly.
- **Social orientation and communication style.** For each task, ask yourself if it requires the employee to work more with *people* or with *things*. If a given task requires working more with people, then label it "E>S." If the task requires more working with things, label it "S>E."
- **Work pace.** If a task requires a fast, urgent pace, then label it "A>P" ("assertiveness over pace"). If the task requires a slower, more methodical pace, then label it "P>A."

After you've labeled each task within a category, add up the total of each "trait identifier," which you'd just created. In other words, add a "total" number for each of the following:

Some traits might fall right in the middle; for example, you might determine that a certain task is best suited to a trait that's neither A>S nor S>A, but rather S=A. (Or it could be E=S or A=P.) That's fine. Add up those equals, as well:

Behavior Scoring Grid

(A) > (S)   _____

(S) > (A)   _____

(E) > (S)   _____

(S) > (E)   _____

(A) > (P)   _____

(P) > (A)   _____

(A) = (S)   _____

(E) = (S)   _____

(A) = (P)   _____

Total   _____

Now, like a codebreaker, you'll start to see "patterns in the numbers." Specifically, *the traits with the highest "total number" for them become your list of ideal or required behaviors for that category.*

Now repeat this process for each "Task" category. Then add up the totals for *all* of the "Task" categories, as shown in the following **"director of sales" example**:

Determining the 3 Primary Behaviors

Task One:
Hiring and
Onboarding

(A) > (S)   (E) > (S)   (A) > (P)

Task Two:
Sales Training

(S) > (A)   (E) > (S)   (P) > (A)

Task Three:
Performance
Coaching

(A) > (S)   (E) > (S)   (A) > (P)

Task Four:
Sales
Administration
and Traing

(S) > (A)   (S) > (E)   (A) = (P)

Totals

2 (A) > (S)   3 (E) > (S)   2 (A) > (P)

2 (S) > (A)   1 (S) > (E)   1 (P) > (A)

0 (A) = (S)   0 (E) = (S)   1 (A) = (P)

Note that any "equal" (" = ") scores should be paired with traits with the majority.

Remember: The behaviors with the highest "total number" become ideal or required traits for that category. So just look at the highest-scoring traits, above, to see that "ideal" list. Then compare these results to the MPO Job Profile for accuracy.

Step Five: Calculate the tasks' percentage times

What's the percentage of time, each day, that the employee should spend on any of the given tasks in your P3 list? We include this percentage for two reasons:

First, you want to gauge the alignment (or lack thereof) with the individual's hardwired traits, to see if these tasks will present a natural way of thinking and behaving for that person; in other words, you want to determine precisely where they may need to "stretch their exercise bands."

Second, you want to provide the candidate with an idea of how much time they may spend doing these exact tasks. This will let them decide if they want to do the job.

Alignment of required behavior and traits is crucial.

Have you ever been told, or told someone, "I've seen you do this perfectly a dozen times; why don't you do it like that all the time"? If that person has, indeed, performed the task flawlessly a dozen times, then they clearly possess the first two strengths that characterize a consistent, productive employee: 1) Results and 2) Knowledge. But in this "why-can't-you-do-it-all-the-time" situation, they're likely missing that final strength: 3) alignment of their traits with the required behavior of the job.

The more responsibility that a role requires, the more behavioral adjustments are needed. As long as the time required to perform a given task is less than 30 percent of the total job, then you needn't be too concerned about the alignment of traits and required behavior—so long as the person can demonstrate that they've done the tasks before, and possess the required knowledge and skills which will let them achieve the required results.

On the flip side, if the tasks require *more* than 30 percent of their time, and if their traits aren't aligned with the job's required behaviors, they can have all the knowledge and skills in the world, but they'll have to exert a ton of effort, "stretching those bands." They'll need to be acutely self-aware of their need to modify their behavior. They'll need to focus on the knowledge required to adjust that behavior. And then, on top of all that, they'll need to expend the effort necessary to sustain the required behavioral change, just to accomplish those tasks.

That's a lot of work.

Not surprisingly, there's a strong probability that that they'll end up inconsistent in their performance of these tasks. They certainly won't enjoy doing them. And they'll eventually either burn out or quit. Or—just as bad—their boss will burn out, given their inconsistent performance, and fire them.

This is why it's so important to align the required behavior of the job with the natural traits of the employee. And that's why hiring for *behavior* is a bad idea. In an interview, people behave in the way they think you want them to. Most of the time, they're acting, and then you hire them, based on *the behavior they displayed in the interview*. Later, about a month or two into the job, their *real* behavior begins to surface, and you ask yourself, "Whatever happened to the person I hired?" Well, that wasn't them. It was their "representative" that they sent to get the job.

Putting the P3 to work

The P3, admittedly, takes some effort to create. But once you have one, it provides an unparalleled in-depth view of a job. The work you put into making one will pay off significantly in terms of the results it will help you realize. It will help you assess an employee's knowledge and skills as they relate to the requirements of their role. It can help you create that employee's individual development plan (IDP).

As I've mentioned earlier, the P3 is like a Swiss army knife. It can help you make better hiring decisions by focusing your interview on the three strengths (knowledge, results, and behaviors). It will help the candidate make an informed decision by understanding the full scope of the role. And as a performance-management tool, it will help you focus on where you need to coach the employee to achieve maximum results.

I've saved the best part of the P3 for last. Despite what I just said about the work required to create one... you really don't need to do it! XP3 Talent maintains a library of over 100 roles, with a pre-populated P3 for each, available as part of your subscription to the XP3 web-based app. To check out the titles available, just visit XP3Talent.com.

How to identify performance gaps—and corrective actions

Identifying Developmental Needs

Easiest to change

Knowledge
Doesn't know something.

Skills
Can't do something.

Motivation
Do they care about something?

Behaviors
Can't be something.

Hardest to change

Back in Chapter 3, I introduced you to the knowledge and skills-assessment tool, which helps you gauge an employee's abilities as they relate to the role. It's a great method for identifying specific gaps.

That's good. But the P3 is even better. Look at the graph on the previous page. It ranks the elements for determining a lack of performance from "easiest to change" to "hardest." The P3 will not only help you identify performance gaps; it will also give you clues to possible corrective actions!

Once again, your focus will return to the "three strengths"; here's how you'll use them in this context:

1. **Results.** Review their performance against the P3's "results/success" measurements. If they're not achieving them, go back to #1. If they *do* have the knowledge, but aren't achieving the results, they most likely need coaching on the proper *application* of the knowledge.

2. **Knowledge.** Ask employees questions to demonstrate that they possess the required knowledge. If they can't answer them to your satisfaction, then they most likely require more training.

3. **Behavior.** Review their MPO against the required behaviors of the job, using the MPO Right Match tool to help you

identify the specific gaps. Determine how big the gaps are, whether they span more than a one-point spread (or not). If the gap is less than three points, then they most likely just need some "behavioral" coaching, *e.g.*, "Be more assertive," "Be more social," "Be more patient," etc. If the gap is greater than three points, and the estimated percentage of time spent on the task is greater than 30 percent, you'll need to confront the reality that the related tasks may not be in their best interest to do. Can you redefine their P3 to eliminate those tasks and replace them with others that better match their talents? Perhaps you can delegate or trade some of those tasks with another employee who's a better fit for those tasks.

The keys to enhanced communication and motivation

If you learn only one thing from this book, it is this: *Communication is most likely the root of most of your challenges in life.* However, if you learn to communicate correctly, it can be the solution to those challenges!

Let's explore this in more detail.

Effective communication: The art of understanding and being understood

What, exactly, is "effective communication"? If you ask ten different people, you'll likely get six or seven different definitions. That in itself is a clue: *Effective communication is very subjective.*

Or is it? This brings us back to the science of talent management: Our brains are hard-wired to communicate in a particular way. And here's the part you need to be cautious about: Whichever way *you* are hard-wired to communicate is how you most likely define "effective communication." It naturally follows then, that at some level, we all believe that everyone else who communicates differently than we do, *is doing it wrong*.

Are they? The truth, as always, is nuanced. There are, it turns out, multiple ways to communicate effectively.

My goal is to provide you with practical strategies (and tools) to improve the effectiveness of your communication by developing your awareness of the *impact* of your own interaction dynamics on others.

How do you do this? First, take another look at the "leadership staircase" I discussed at the beginning of the chapter. To improve the effectiveness of your communication, you need to be self-aware, and then able to self-manage. With the proper adjustments in your communication style, you will be able to influence others.

The error of fundamental attribution

Consider this "chain of command":

- We judge ourselves based on our intentions.
- We judge others based on their observable behaviors.
- We then process their observable behaviors through our own filters and assign motive.

Thus the "fundamental attribution error" is our innate tendency to explain someone else's behavior based on *internal* factors (such as personality or disposition), while underestimating the influence and importance of *external* factors (such as situational influences) have on a person's behavior.

Here's an example: A person might pursue a line of questioning with the sincere intent of fully understanding an idea that's being presented to them. Sounds fair and reasonable, right? But it's not always *perceived* that way. The person who's presenting the idea might falsely assume that the other person is:

- Questioning the *quality* of their idea.
- Assuming that the idea is wrong.
- Simply seeking an argument.

None of these is true. The questioner's intent was simply to understand the idea.

Sure, there are times when we're correct about our assumptions, but the fundamental attribution error is our tendency to explain the behavior of others based on their character or disposition. Given the diversity of communication styles, the probability that your assumptions will be correct is less than 25 percent. This is particularly true when the behavior is negative. Given the scenario above, how would the false assumptions impact future conversations between the two individuals?

The simple solution

You can assign people's communication style to one of two camps: introverts and extroverts:

- Introverts' communication tends to be more serious, factual, and direct.
- Extroverts' communication style tends to be more open, friendly, and persuasive.

To make the proper adjustment, simply notice the style of the person you're speaking to, utilizing the above descriptors. Then adjust your style to match that of the person you're communicating with.

The complete solution

Clearly, communication is a much deeper subject than what I've described above. While "the simple solution" is a good start, you're going to need to take a deeper dive if you really want to be an effective communicator.

To help you with this, we will take advantage of another tool available through Ngenio called the MPO Communication Report.™ This will help you to pinpoint, with laser-like precision, a person's communication style and interactive dynamics.

The MPO Communication Model™

This model is based on four distinct communication styles:

1. **Assertiveness and self-control.** This is the *authoritative*, centered on results.
2. **Assertiveness and spontaneity.** This is the *expressive*, centered on people.

3. **Cautiousness and spontaneity.** This is the *cooperative*, centered on consensus.
4. **Cautiousness and self-control.** This is the *analytical*, centered on tasks.

Nuances in workplace communication styles

The MPO Communication Model incorporates the above styles into a four-quadrant matrix (see graphic on the next page), taking into consideration the fact that behavioral variations occur within each of the four quadrants. These variations (called "nuanced styles") are based on the particular position an individual resides along the two axes of "Assertiveness/Cautiousness" and "Spontaneity/Self-Control." Each quadrant has been subdivided into four segments in order to reflect specific behavioral subtleties based on a person's positioning along the axis.

MPO Communication Model

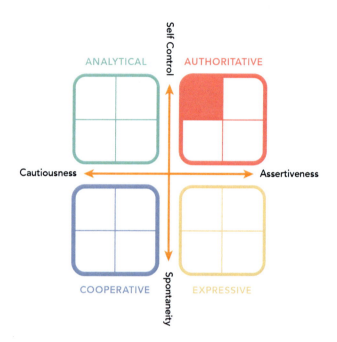

The MPO Communication Report helps anyone to clearly understand their unique communication style and most importantly, how they are perceived by their co-workers based on which quadrant they fall into. It then provides specific coaching recommendations on how to adjust one's communication to be the most effective, based on their style.

To give you an example, the above graph is an excerpt from an MPO Communication Report of someone who is in the upper left quadrant of an "Authoritative." Here's how that section of the report reads:

GENERAL CHARACTERISTICS

You enjoy being in control of a situation and prefer things being done your way and at once. You aim to get results and reach your objectives while maintaining sound relationships with others. You rely on your own opinions and generally keep your feelings and motives to yourself. What's more, you are not afraid to stand out from the pack and take a position. You like to be viewed as having a strong and reliable character.

You are naturally:

- *Competitive and like to take matters into your own hands*
- *Directive and persuasive; your idea of how things should be accomplished is firm and fixed*
- *Goal-oriented, highly organized and planned*
- *An intense, passionate and sociable leader*

Your main assets:

- *Dynamism, energy and drive*
- *Initiative, ability to seize opportunities*
- *Efficiency and determination toward goals, not easily discouraged by obstacles*
- *A fighting spirit, not deterred by problems and obstacles*
- *An ability to combine audacity with cautiousness*
- *Sense of urgency and the ability to deliver on tight deadlines*

Others, however, may perceive you as someone who:

- *Is impatient and inclined to be highly critical*
- *Needs to be right (beware of your pride)*
- *May procrastinate on decision-making when under stress*
- *Imposes own ideas to achieve personal objectives*
- *Sometimes does not listen to others whose opinions differ from their own*
- *Controlling and overly picky when it comes to important details*

Although you prefer being a lone-ranger, you make decisions and plan while taking time to inform others in order to gather support for your projects. As you always focus your energy on achieving results, you tend to impose your views upon others and only delegate sparingly. When you face resistance, you instinctively become more insistent, adopt an authoritarian attitude and become less communicative, unless the arguments put forth seem solidly structured.

The MPO Communication Report then goes on to provide you with *specific suggestions for enhancing your effectiveness with each of the other quadrants;* how cool is that? Here's an excerpt from that section of the report:

DEALING WITH COOPERATIVE TYPES

This profile is the furthest from your own and the one you have the most trouble getting along with. From your point of view, Cooperative personalities lack initiative and stick too closely to the status quo, rather than accepting change. Nevertheless, this may be complementary to your profile if you know how to make the best of your relationship:

- *Slow down your pace*
- *Take time to establish common ground*
- *Show interest in the person*
- *Demonstrate that you understand their perceptions*
- *Display interest in what the person does*
- *Clearly explain how you see things and what you expect from the Cooperative*

- *Satisfy the person's need for support by recognizing efforts and accomplishments made*
- *Guide and direct the Cooperative when facing obstacles*

As you can see, there is a science to effective communication and the MPO has figured it out for you! With that "teaser" in mind for creating your own Communication Report, here are some overall guidelines that you can put to work today to enhance the effectiveness of your communication:

- Learn more about your natural communication style.
- Learn the communication styles of your co-workers.
- Lose your judgments around what's "normal" and "correct." Instead, appreciate the differences—and understand that both styles can both work effectively.
- Adjust your communication style to accommodate the style of the person with whom you're communicating.

Understanding (and leveraging) behavioral motivation

Everything we do is related to our desire to satisfy

a fundamental need. Psychologist Abraham Maslow famously described this in his five-tier "Hierarchy of Needs."

But not everyone's needs are the same. This seems to take The Golden Rule—the precept which says that we should treat others the way we wish to be treated—and stand it on its head. Everyone has their own, unique genetic code, and thus their own set of needs—and optimal way to be treated. Therefore, to be a highly effective performance coach, you need to learn the unique needs of each of your employees. You've got to live by the *Platinum* Rule, which states: "Treat others the way *they* want to be treated"!

Maslow himself had a great quip about this, too: "If the only tool you have is a hammer, you tend to see every problem as a nail."

External vs. internal motivators

To motivate your employees, there are two areas where you could focus: external motivators, and internal motivators. Let's compare and contrast them:

External motivators include things like "I want a new car, house, boat, etc." As a manager, it's difficult for you to influence these external motivators. You could, say, encourage and focus your employees on making more money, but even if you're successful, these external motivators are short-lived. You'd soon have to find the next "thing" that will motivate your employee. And you also need to be aware that, many times, when people desire certain "things," those "things" can actually be false motivators— masking the true *internal* motivating factor such as status symbols for fulfilling their ego.

In contrast, a much easier and longer-lasting method of motivating your employees returns us to the science of behavior. Remember, everything we do is related to our desire to satisfy a fundamental internal need. Here's a sampling of those intrinsic motivational needs, categorized by the four MPO personality types:

- **High As** are motivated by independence, control, challenges, and opportunities to win.
- **High Es** are motivated by recognition, praise, image, and inclusion.
- **High Ps** are motivated by stability, process, consistency/routines, and familiarity
- **High Ss** are motivated by clear expectations, structure, and performance feedback.

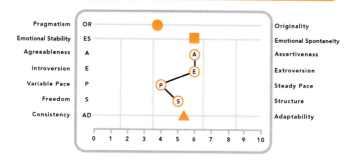

Decoding what they say—and what they mean

Once you understand the science of these intrinsic motivators, it becomes almost painfully easy to figure out what these different personality types are *really* saying, when they tell you something; it's like you can see closed-captioning of their "translated speech." Let's just take "I want to make more money" as an example:

- When a High A says they want to make more money, what they really mean is that they want independence and control.

The more money they have, the more independent they are, and the greater the control they have over their life.

- When a High E says they want more money, it's usually to satisfy their intrinsic need for image and recognition.
- When a High P says they want more money, they're typically satisfying their need for stability.
- When a High S says they want more money, they seek to satisfy a need for recognition of the quality work they do.

Once you understand the intrinsic motivational needs of each of your employees, you'll then be able to create environments that provide continuous motivation which meets their needs at a deeper level. Consider these examples:

- Provide your High A's with as much independence and autonomy as possible. To meet their need for control, give them a project, territory, team, etc. that they can be in charge of. Conversely, don't put them in roles that fail to provide them with independence, or they won't stick around long. To a High A, that's a de-motivator.
- To motivate your High E's, you'll need to communicate with them regularly in a manner that shows you care about them.

Provide them with public recognition so everyone else knows when they do a good job. Conversely, *don't* put them in jobs where they won't get the opportunity to regularly interact with others. Remember: Their vitality comes from social interaction. Without that, their energy will be quickly sapped—along with their productivity.

- High P's need a stable, consistent environment. They like and/or need their routines. They require a slower pace. Create as stable and consistent of an environment as possible for them. Give them as much time as possible to get things done. Be sure to thoroughly discuss any change, and create a plan to implement them. To avoid de-motivating them, don't put them in roles where there is constant change or too fast of a pace.

- To motivate a High S, provide them with constant feedback about the quality of the work they produce. Recognize them for their ability to organize things, maintain order, and for being compliant and ensuring that others are also being compliant. To them, it's only fair. De-motivators include lack of clarity, disorganization, or taking a "close is good enough" approach to things. They want and need precision.

Effective one-to-one (121) guidelines

[handwritten: Forget all this + use Tools Manager model + advice for kids]

Why should you do a one-to-one, or "121," with your employees—especially when you're already talking to them every day? That's the question I'm perennially asked by managers, when I ask them whether they do 121s with their staff. The reason managers don't believe they need to conduct a formal one-to-one is because they don't understand its purpose, nor how to conduct an effective one.

To help you understand what a 121 is, let's start with what it *isn't*. A 121 is not a time to discuss the routine issues you cover every day. It's not a "status report." Nor is it a tactical meeting to develop a new idea or plan. Without a proper framework and mindset, your 121 can degenerate into just another meeting in your day. Thus the following are suggestions for making your 121 one of your most important meetings for both you and your employee.

What a 121 meeting is for

One of the most important elements of a successful 121 is the creation of a safe space wherein your employees feel comfortable to discuss the issues and concerns on their mind. Done properly, 121s ensure that 1) your employees are in alignment with their objectives, 2) are focused and working on those goals, and 3) that you are there to support them with those accomplishments. One-to-one's are for regular check-ins to avoid larger issues from festering; they allow for immediate feedback and promote open communication. The 121 exists to help you, and the employee, to discuss the specific details of a plan to accomplish immediate objectives, any challenges or roadblocks they are having, how they will overcome them, and what support they need from their supervisor to accomplish those objectives.

Some of the benefits of an effective 121 meeting

1. They strengthen the relationships between the manager and their team members.
2. They improve productivity by creating focus and alignment.
3. They are a place for coaching, mentorship, focus, and alignment.

121 responsibilities

- The *employee's* role is to take responsibility for what they need from you as their supervisor.

- *Your* responsibility is to provide support and add value that helps the employee perform well.

The 121 schedule

Stick to a schedule. Set aside 30 to 60 minutes with each of your team members on a weekly, bi-weekly, or monthly basis.

And don't feel you need to always conduct the 121 in a formal setting. Occasionally, I suggest getting out of the office: Go for a walk or grab a coffee. This helps create a more relaxed environment that facilitates an open dialog.

How frequently should you conduct 121s?

The answer to this question will vary depending on the role of the employee, the volume of items the employee is working on, and the urgency of those items. Use these guidelines:

- 121s should be held weekly for an employee with a lot of urgent items such as a salesperson.
- Most can be done biweekly.

- For other roles that do not have a lot of change or urgency, they can be conducted monthly.

All that said, the primary objective is to be *proactive*. If you are not meeting at least once per month, you are being *reactive*. Once you get past 30 days, it's really hard to play catch-up.

The 121 meeting structure

Your employees should come prepared to answer the following three questions, keyed to each of their current objectives:

1. What have you accomplished on this since our last one-to-one?
2. What is your plan to accomplish this objective between now and our next one-to-one?
3. What can I do to help you with this objective?

That third question is important. "Help" can come in the form of teaching and coaching them on how to do it. It can mean providing them with resources, tools, people, time, funding, and so forth. It does not mean that you will do the job for your employees!

Congratulations!

In this chapter alone, you've learned about performance coaching, the leadership staircase, and the journey to self-awareness. You've learned all about the powerful P3 and how to put it to use. You've discovered the keys to enhanced communication and motivation. And you've learned how to conduct effective one-to-one's.

Which might make you ask: What's next?

Turn the page and find out.

NEXT STEPS

Three things
you can do right now

Thank you for reading *Hire • Train • Retain*. By completing this book, you now know more about those three topics than most people, at most companies. To jump-start your success, and benefit immediately from the ideas in this book, here are three things you can do, right now:

1. **Have your leadership team and employees read this book.** This will give everyone a common language and context for implementation.

2. **Create an implementation team that will develop your plan for execution.** This plan should include what will be implemented, how it will be implemented, who will do what, and by when. Then

ule frequent (no less than monthly, ideally biweekly) review/strategy ngs to ensure success.

. ...ment a quarterly theme. Depending on the size of your team and available resources, you may not want to try and tackle everything at once. Choose a topic—"Hire," "Train," or "Retain"—for the quarter. Set some goals for what you intend to implement. Then focus the entire company on achieving those goals.

Free 60-day Trial of the XP3 web-based App

As my gift to you for taking the time to read my book, I am offering a free 60-day trial to the XP3 Recruiting App. To take advantage, send an email to XP3demo@XP3talent.com

Sharpen your skills

If you're the extra-motivated type, I'll bet you'd love to get your hands on even more information and tools. If you'd like to become a true master of this discipline, with skills at the expert level, there's even more to the XP3 Talent System than this book, our website, and app. We also offer the outstanding XP3 Leadership Academy, where you can immerse yourself in workshops and exercises, and truly attain mastery.

The XP3 Leadership Academy

When you and your management team attend the XP3 Leadership Academy, you'll get the training you need to acquire the fundamental knowledge—and a shared language with other managers to implement the process and systems to help you grow.

But training, by itself, almost never works. Training is just information; it's a foundation for building skills. To truly succeed, people must *use* these skills and force out older habits. Thus, the XP3 Leadership Academy is a comprehensive five-day program that can also be divided into four different sessions, in order to accommodate your team's needs. It also includes monthly webinars and individual implementation coaching sessions for 12 months.

For participants who haven't read this book, the XP3 Leadership Academy gives them this information. And for readers of this book, like you, you'll get much more. You'll learn how to:

- **Use analytical tools.** All attendees of the XP3 Leadership Academy get subscriptions to both the XP3 Talent and MPO apps, so you'll have all the analytical tools you need, right at your fingertips. You'll learn about the Ngenio MPO behavioral suite of tools and the various functions and characteristics of the MPO. These include the MPO Personality Profile, the MPO Job Profile, Right Match, Talent, and Communication. All of these will be utilized in Academy workshops as part of your "talent tool kit." In hands-on sessions, you'll learn how to interpret the graphic and text results of an MPO to extract insights—the way medical professionals read an EKG—to fast-track your "diagnosis and treatment plan." Working with teammates, you'll learn to read the reports of yourself and your peers. You'll practice interpreting MPO graphs and application techniques—and gain insight into each other's profiles, communication styles, and motivational needs. You'll get a full set of manuals, reference books, cheat sheets/quick reference guides, and templates. You'll even get a framed certificate, denoting your attainment of the title of "Certified MPO

Behavioral Analyst," upon completion of the XP3/MPO 40-hour leadership course.

• **Build real-world skills.** You'll get to apply everything you've learned in the book to yourself, your teammates, and your company. Working with XP3 Talent instructors and your teammates, you'll learn how to interview like a pro—with practice role-playing of interviewing skills and techniques. You'll learn how to develop and implement a structured, foolproof onboarding training program. Using your MPO knowledge and actual report results, you'll discuss issues such as underperformance, poor communication, inability to get along with others, and lack of motivation. Then you'll learn the specific methods you need to address those more effectively back on the job. While you learned about the three levels of engagement in this book, the Academy will help you learn how to identify and address each level—and then how to elevate more of your people to the proper engagement level, and why this is so critical for them and your company's growth. You'll take the knowledge-and-skills template shown in the book—and walk away with a completed assessment for one of your roles and employees that they can use to become more effective. You'll leave with specific plans and tools for your employees and your company.

- **Gain year-long learning.** While this book has taught you how to implement a simple, yet highly effective, hiring process, the Academy goes much deeper. It includes the XP3 Recruiting App subscription—and teaches you how to utilize it. Most subscription-based applications require you to learn on your own, or perhaps attend a few webinars. By contrast, the Academy includes 40 hours of intense learning, so you can return to the office with an expert level of application. Even better, the Academy is not limited to the 40 hours you'll spend in-session. Rather, it's part of a year-long learning and coaching program, wherein XP3 Consultants are available to guide, coach, and enhance the application of the lessons learned in the Academy. For each participant, this includes four private coaching sessions, webinars, and unlimited phone support for all things that were covered in the Academy, from your XP3 Coach who is your own private subject-matter expert, helping you to expand your learning and application. In fact, to ensure that all XP3 Talent systems, processes, and training deliver the desired ROI, we provide coaching that hand-holds each key individual through the required changes to ensure success, helping to identify and develop a focused approach on the necessary skills required to become highly effective.

To learn more about the XP3 Leadership Academy, and all of our other offerings, simply visit XP3Talent.com.

ABOUT THE AUTHOR

Mike Matalone

 Founder of the XP3 Talent System and Leadership Academy.

An award-winning speaker and nationally recognized "talent guru," Mike Matalone is the Founder and Chief Talent Officer of the XP3 Talent System and Leadership Academy. A highly-sought-after trainer, workshop leader, and keynote speaker, he's published 100's of articles on the subject. Mike's engaging personality and uncanny ability to translate tricky concepts into easy-to- grasp "nuggets" of information have made him a prized addition to any talent initiative; in the past 17-plus years, he's delivered more than 500 workshops and keynote presentations to business groups throughout the United States, Canada, and the U.K.

Started 4/4/20
Finished 4/16/20